T0133903

Machine Learning and its Applications

Peter Wlodarczak

University of Southern Queensland
Toowoomba, Queensland, Australia

CRC Press
Taylor & Francis Group
Boca Raton   London   New York

CRC Press is an imprint of the
Taylor & Francis Group, an **informa** business

A SCIENCE PUBLISHERS BOOK

CRC Press
Taylor & Francis Group
6000 Broken Sound Parkway NW, Suite 300
Boca Raton, FL 33487-2742

Printed on acid-free paper
Version Date: 20190821

International Standard Book Number-13: 978-1-138-32822-8 (Hardback)

Library of Congress Cataloging-in-Publication Data

Names: Wlodarczak, Peter, author.
Title: Machine learning and its applications / Peter Wlodarczak, University of Southern Queensland, Toowoomba, Queensland, Australia.
Description: First edition. | Boca Raton, FL : CRC Press/Taylor & Francis Group, [2020] | Includes bibliographical references and index. | Summary: "This book describes Machine Learning techniques and algorithms that have been used in recent real-world application. It provides an introduction to Machine Learning, describes the most widely used techniques and methods. It also covers Deep Learning and related areas such as function approximation or. The book gives real world examples where Machine Learning techniques are applied and describes the basic math and the commonly used learning techniques"-- Provided by publisher.
Identifiers: LCCN 2019033419 | ISBN 9781138328228 (hardcover ; acid-free paper)
Subjects: LCSH: Machine learning.
Classification: LCC Q325.5 .W635 2020 | DDC 006.3/1--dc23
LC record available at https://lccn.loc.gov/2019033419

Visit the Taylor & Francis Web site at
http://www.taylorandfrancis.com

and the CRC Press Web site at
http://www.crcpress.com

For my parents

Preface

Artificial intelligence or AI, and in particular, machine learning has received a lot of attention in media due to the rapid advances we have seen in this area over the past few years. Progress in artificial intelligence has been slow in the 1990's and a certain disillusionment was noticeable in research since artificial intelligence did not seem to live up to the expectation. Since the beginning of the new century, machine learning has seen rapid progress due to new approaches, in particular, deep learning. Deep learning goes further back, however, due to recent advances in other areas, in particular, the development of graphics processing units that provide the necessary computing power, and the Internet, that made an unprecedented amount of data publicly available, machine learning has gained a lot of momentum and we have seen an incredible amount of new applications in many areas of research and in practice. Machine learning is used for spam filtering, for medical image analysis, for voice commands, for autonomously driving cars and to analyze sensor data for the Internet of Things just to name a few. Every high-end smart phone is now AI enabled. Deep learning is used to predict earthquakes, for automatic language translation, for automatic coloring of black and white movies and for forecasting in finance. The Mars rover Curiosity utilizes artificial intelligence to autonomously select inspection-worthy soil and rock samples with high accuracy. The list goes on.

Recently, machine learning also attracted a lot of attention due to visions of jobs being lost to machines, dangers of autonomous systems, such as robots who could turn evil, loss of control over existential infrastructure or even a war starting due to artificial intelligence. However, it should be noted that artificial intelligence, as the name suggests, is artificial and has nothing to do with human intelligence. A machine may be capable of recognizing whether an animal in a picture is a lion or a tiger, but it is not able to understand the concept behind what it recognized, i.e., it does not understand the concept of a living being. A deep learner is, in essence, a huge mathematical formula. It has nothing to do with how the human brain works, where biochemical processes are executed. An artificial neural network is inspired by nature. It has a grain of how we think the brain works and a lot of math thrown at it. Visions of machines replacing doctors in diagnosing a patient or lawyers in sentencing a perpetrator are probably premature. It is more likely that we will see artificial intelligence supporting a doctor or a lawyer in his daily work instead of replacing her or him in the near future.

However, we will likely see many more applications of artificial intelligence in areas that we have not thought of yet. Artificial intelligence and machine learning pose many interesting problems in practice and in research. The fact that we try to imitate nature gives us a better understanding of what problems nature had to solve during evolution. It also helps us to better understand how it solved them. However, how human learning works is still largely unknown. Before we have a better understanding of how the brain works, it is unlikely that we will be able to replace it, no matter how much math we apply.

When I started to work with machine learning there were excellent books explaining the math used in machine learning algorithms. There were also excellent books about data mining and knowledge discovery. Some of them are listed in the reference section. They can be used for studying and for reference. However, there were no books explaining the basic concepts behind machine learning and how they are applied in practice. It is important to understand the math and the algorithms used in machine learning. However, it is not necessary to know all the details about the algorithms to be able to effectively use them for real world problems. Machine learning has evolved over the past years to the point where a lot of the complexity is hidden in frameworks such as TensorFlow or Apache Mahout. A data scientist can use these frameworks without knowing the implementation details. Many popular programming languages have libraries with implementations that are ready to be used in programs. R, Python, Java or Scala have many different implementations of machine learning methods and the data scientist can continue using his preferred programming language by including the libraries that fit the purpose best. There are also tools, such as RapidMiner, Weka or Knime to name a few, where no programming skills are required and machine learning workflows can be created graphically. Obviously, next to the open source implementations, there are also commercial tools offered by different vendors to choose from.

This book aims to explain the basic concepts behind machine learning and the machine learning methods. It is the work of several years of experience in applying these methods in practice in various projects. It tries to give a concise description of the algorithms and the math behind them to the level where it helps explain the inner workings. However, it is not intended to give an exhaustive mathematical description with all the derivations. Also, machine learning goes back to the 1940's and has evolved since then and many different techniques have been developed, too many to describe in one single book. Also, for virtually every machine learning method, variations have been proposed that might be more suitable for certain problems than others. Nevertheless, some of the basic concepts apply to many of these methods and understanding them makes it easy to familiarize oneself with new methods that have not been used before. This book is written in such a way that each chapter can be read individually with the caveat that some redundancy exists in the chapters. This book will hopefully make it easier for the reader to get started with machine learning and support them in the fascinating journey through the world of data science and machine learning.

Contents

SECTION II: MACHINE LEARNING

SECTION III: DEEP LEARNING

SECTION IV: LEARNING TECHNIQUES

List of Figures

List of Tables

INTRODUCTION I

Chapter 1

Introduction

CONTENTS

Machine Learning (ML) is a sub-area of artificial intelligence, or AI. The central scientific goal of artificial intelligence is to understand the principles that make intelligent behavior possible in natural or artificial systems [31]. Early artificial intelligence research showed a lot of interest in making computers reason and deduce facts [25]. In recent research, the focus of artificial intelligence has shifted towards designing and building agents that act intelligently. An agent can be a living object, such as a human or an animal, or it can be a robot, a sensor or a car. In artificial intelligence, we are interested in computational agents whose decisions can be explained. Artificial intelligence is founded on mathematics, logic, philosophy, probability theory, linguistics, neuroscience, and decision theory. It has many applications in computer vision, robotics, natural language processing (NLP) and machine learning.

The goal of machine learning is to enable computers to learn on their own. Machine learning is about making computers modify or adapt their actions (whether these actions are making predictions, or controlling a robot) so that these actions become more accurate, where accuracy is measured by how well the chosen actions reflect the correct ones [25]. Using machine learning, a computer can learn rules by itself without the need of a programmer to develop them manually. This is particularly useful when the rules get too complex for a developer to implement, or when there are too many rules to be programmed manually. An example where machine learning is used are spam filters. Spam filters classify mails into legitimate and unsolicited mails, i.e., spam. There are constantly new forms of spam appearing and there are simply too many different types of spam mails for it to be possible to program the rules to recognize them manually. Spam filters learn the rules from past spam mails and learn new rules as new types of spam appear. How computers learn to solve problems without being explicitly programmed is the central question this book tries to answer.

Machine learning is nothing new. The term "Machine Learning" was coined by Arthur Samuel in 1959 while working at IBM. He defined machine learning as a field of study that gives computers the ability to learn without being explicitly programmed [2]. In his seminal work "Some Studies in machine learning Using the Game of Checkers" [35], he probably describes the first program with the capability to learn. His checkers program learned to improve its playing by itself. A computer can be programmed so that it will learn to play a better game of checkers than can be played by the person who wrote the program [35]. However, machine learning goes further back, to the 1940s, with McCulloch and Pitts' theories of biological learning [26] and Rosenblatt's perceptron, the first artificial neuron, described in 1958 [34].

Machine learning has gained momentum in the past years due to big progresses, in particular using deep learners, that have been made in object and speech recognition and autonomous systems, such as autonomously driving cars, automatic machine translation, image caption generation or coloring black and white images.

Machine learning is an interdisciplinary area of research based on statistics, probability, neuroscience, psychology and physics. Machine learning is a mathematical formalization of learning. As such, there is a discrepancy between natural learning in humans and animals, which is a biochemical process, and learning on computers, which is a mathematical abstraction. In humans, learning is carried out in wetware, in computers it is carried out in hardware.

1.1 Data mining

Data mining (DM) is defined as the process of discovering patterns in data [40]. There is no clear distinction in literature between data mining and machine learning. In some publications, data mining focuses on extracting data patterns and finding relationships between data [10], whereas machine learning focuses on making predictions [36]. Data mining is the process of discovering useful patterns and trends in large data sets. Predictive analytics is the process of extracting information from large data sets in order to make predictions and estimates about future outcomes [17]. Therefore the distinction is in the aim. Data mining aims to interpret data, to find patterns that can explain some phenomenon. Machine learning on the other hand, aims to make predictions by building models that can foresee some future outcome. However, clustering is a machine learning technique that aims to understand the underlying structure of the data and has similar goals to data mining. Here, we treat machine learning as a subarea of data mining where the rules are learned automatically.

Humans learn from experience, machines learn from data. Data is the starting point for all machine learning projects. Machine learning techniques learn the rules from historic data in order to create an inner representation, an abstraction, that is often difficult to interpret. Programming computers to learn from experience should eventually eliminate the need for much of this detailed programming effort [35].

1.2 Data mining steps

A typical data mining workflow goes through several steps. These steps include data collection, cleansing, transformation, aggregation, modeling, predictive analysis, visualization and dissemination. However, depending on the problem at hand, those steps might vary or additional steps might be necessary. Data mining requires domain specific knowledge. If a data mining project is supposed to detect fraud and money laundering in financial transactions or group news articles into categories such as "Politics", "Business", "Science" etc., different techniques apply. Getting familiar with the domain of the application and setting the goals of the data mining project are necessary preliminary tasks in order for the data mining project to complete successfully. Domain knowledge is also necessary to evaluate the performance of a trained learner. Figure 1.1 summarizes the data mining steps.

Figure 1.1: Data mining steps.

Data mining is a highly iterative process and typically goes through many iterations until satisfactory results are achieved. Data mining projects have to be executed with great care. The growth of data in the past years has fostered the development of new data mining applications where the internal workings often go undocumented. The black-box approach is dangerous since the results can be difficult to explain, or lead to erroneous conclusions. For instance, cancer tissue in medical images that was undetected or a traffic situation that was misinterpreted by an autonomously driving car. The ease with which these applications can manipulate data, combined with the power of the formidable data mining algorithms embedded in the black-box software, make their misuse proportionally more hazardous [17]. Ultimately, one can find anything in data and if a machine learning project is carried out without proficiency it can lead to expensive failures.

1.3 Data collection

Data collection is the process of tapping into data sources and extracting the data needed for training and testing a model. Data sources include databases, data warehouses, transaction data, the Internet, sensor data from the Internet of Things or streaming data, but many more sources exist. If the data is stored in its original format, it is called a data lake. The data can be historic or real-time, streaming data. For instance, training a model for spam filtering needs historic, labeled legitimate and spam mails for training and testing. Spam mails are artfully crafted in order to avoid elimination by spam filters, making spam filtering a tricky task. Spam filtering is one of the most widely-used applications of machine learning.

Often there is more than one data source and multiple data sources need to be combined, a process called data integration. As a general technology, data mining can be applied to any kind of data as long as the data are meaningful for a target application [10].

It is often difficult to collect enough training data. Data is sometimes not publicly available or cannot be accessed for privacy or security reasons. To mitigate the problem of sparse data sets, synthetic data can be created, or training techniques such as cross-validation can be applied. Also, to train a learner, labeled data is needed. Raw data, such as emails, are not labeled, and producing labeled training sets can be a laborious task. Semi-supervised techniques such as active learning can be used in these situations. Active learning is a form of online learning in which the agent acts to acquire useful examples from which to learn [31]. In offline learning, all training data is available beforehand, whereas in online learning the training data arrive while the learner is trained. The learning algorithm can actively ask the agent to label data while it arrives. Data availability, whether labeled or not, is crucial for the success of a machine learning project and should be clarified before a machine learning project is initiated.

1.4 Data pre-processing

A problem that plagues practical machine learning is poor quality of the data [40]. Real world data is often noisy and inconsistent and cannot be used as is for practical machine learning applications. Also, real world data is seldom in a format that can be directly used as input for a machine learning scheme. That is why data pre-processing is needed. Data pre-processing is usually where most of the time is spent. It is not unusual that it takes up to 70% of the effort in a data mining project.

Whereas machine learning techniques are domain independent, data pre-processing is highly domain specific. For instance, depending on whether text data is analyzed or images, different pre-processing steps apply. The pre-processing tasks also depend on the learning algorithm applied. Some algorithms can handle noise better than others. For instance, linear regression is very sensitive to outliers, which makes outlier removal a mandatory pre-processing step.

There are many different pre-processing techniques. Typical data pre-processing tasks next to outlier removal include relevance filtering, data deduplication, data transformation, entity resolution and data enrichment. For instance, going back to the spam filter example, spam mails typically contain words or phrases such as "buy online", "online pharmacy" or hyperlinks more often than legitimate mails. The frequency of certain words or phrases gives an indication of whether the mail is spam or legitimate. Irrelevant words, characters or symbols are first removed from the mail, a process called stop word removal. There is no authoritative list of stop words and they depend on what is being mined for. Stop words are the most common words in a language, e.g., "the", "who", "that". After stop word removal, the frequencies of the remaining words are counted, to create a word list with their frequencies. The resulting list is called a bag-of-words. The bag-of-words is then used as input for a machine learning scheme. Figure 1.2 shows an email before and after pre-processing.

Since machine learning algorithms usually do not take text as input, creating a bag-of-words with their frequencies is a very common pre-processing step in many text analysis tasks. It is simple but effective

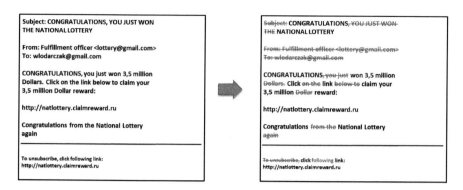

Figure 1.2: Email pre-processing.

for tasks such as opinion mining, text classification or information retrieval.

Extracting the relevant information from a data set is called feature extraction. It selects the relevant observation points from raw data. Sometimes the feature set is reduced, a process called feature selection. Feature selection helps to avoid overfitting since it reduces the complexity, but it should still describe the data with sufficient accuracy. The resulting set of features is represented as a feature vector. A feature vector is what many machine learning algorithms use as input. Features can be numerical, e.g., the age of a person, or categorical, e.g., the job title of a person. The input for a machine learning algorithm can be a tensor with any number of dimensions. If the tensor has one dimension it is a vector, if it has two dimensions it is a matrix.

Data transformation is another common data pre-processing step. It converts data into a format that can be used as input for machine learning algorithms. For instance, converting feet into meters, strings into numeric values or continuous into discrete values are common transformation tasks.

In 1997, IBM's Deep Blue defeated chess champion Garry Kasparov. This was hailed as one of the milestones of machine learning. From a pre-processing point of view, chess is a simple task since chess has clearly defined rules and a move such as "Bishop c3 - d4" is unambiguous. In other words, chess data has no noise and a limited set of rules.

1.5 Data analysis

Data analysis is the process of knowledge discovery. During the data analysis phase, the predictive model is created. There are many data analysis methods that do not use machine learning techniques. However, this book focuses on data analysis using machine learning, other methods are beyond the scope of this book.

Machine learning is divided into supervised, semi-supervised and unsupervised learning. Some widely-used learning methods include ensemble learning, reinforcement learning and active learning. Ensemble learning combines several supervised methods to form a stronger learner. Reinforcement learning are reward-based algorithms which learn how to attain a complex objective, the goal. Active learning is a special form of semi-supervised learning.

1.5.1 Supervised learning

Supervised learning techniques are applied when labeled data is present. The labeled data is used for training and testing. Labeling is often a manual process and can be time consuming and expensive. Every training data record is associated with the correct label. For spam filtering, labeled data will mean a data set of spam and of legitimate mails. Here, the labels are "spam" and "legitimate". During training, the machine learning algorithm learns the relationship between the email and the associated label, "spam" or "legitimate". The learned relationship is then used for classifying new emails that the learner has not seen before into their corresponding category.

Supervised methods can be used for classification and regression. Classification groups data into categories. Spam filtering is a classification problem since mails are classified into spam and legitimate mails. The classes are the labels. Since there are two categories, it it a binary classification problem. If there are more than two classes, it is called a multi-class or multi-label classification problem.

Regression analysis is used for estimating the relationship among variables. It tries to determine the strength of the relationship between a series of changing variables, the independent variables, usually denoted by X, and the dependent variable, usually denoted by Y. If there is one dependent variable, it is called simple linear regression, if there is more than one dependent variable, it is called multiple linear regression. In classification, you are looking for a label, in regression for a number. Predicting if it is going to rain tomorrow is a classification problem where the labels are "rainy" or "sunny", predicting how many millimeters it is going to rain is a regression problem. The target or dependent variable y is a continuous variable. Contrarily, discrete variables take on a finite number of values. Typical supervised methods are Bayesian models, artificial neural networks, support vector machines, k-nearest neighbor, regression models and decision tree induction.

1.5.2 *Unsupervised learning*

Unsupervised learning techniques are used when no labeled data is present. In other words, there is no y. One of the most popular unsupervised approaches is clustering. The goal of clustering is to understand the data by finding the underlying structure of data. Clustering groups data based on some similarity measure. For instance, a company groups it's online users into customer groups with similar purchasing behavior and demographics in order to better target them with a marketing campaign. In this example the similarities are the purchasing habits and the age group. Since there is no labeled data, evaluating a cluster is not an obvious task to do. There is no ground truth that the result of a clustering task can be compared with.

Clustering can also be used to represent data in a more compressed format (dimensionality reduction), for data summarization, while keeping it's structure and usefulness. Clustering can also be used as a preliminary step for supervised machine learning algorithms, for instance, to reduce dimensionality to improve performance of the supervised learner.

Typical clustering algorithms include k-means clustering, hierarchical clustering or principal component analysis (PCA). In k-means clustering, the algorithm groups data points into k groups where k is the center of a group and is called centroid. The centroid is also called geometric center or barycenter and represents the mean position of all the data points of a cluster group. In k-means clustering, the data points are grouped around the centroids. The points closest to the centroid k are added to the cluster. For instance, a warehouse inventory is grouped by sales activity or sensor data in an Internet of Things (IoT) application is grouped into normal and deviant sensor data for anomaly detection.

In k-means clustering, the number of k has to be defined beforehand. Selecting the number of clusters is not always an obvious task to do. For instance, grouping images of animals into their biological families such as felines, canines, etc., requires prior knowledge about how many families there are going to be in the images to be clustered. k-means is easy and converges quickly, which often makes it a good starting point for a machine learning project.

Hierarchical clustering or hierarchical cluster analysis (HCA), groups data by creating a tree or dendrogram. There are two approaches, a bottom up or agglomerative approach, where each data point starts in it's own cluster, and the top down or divisive approach, where all observations are put into one cluster and splits are performed in order to create a hierarchy that is usually presented as a dendrogram.

Human and animal learning is largely unsupervised. Humans learn from observing the world, not by being told the label of every object, making unsupervised learning more biologically plausible.

1.5.3 Semi-supervised learning

Semi-supervised learning is typically used when a small amount of labelled data and a large amount of unlabelled data is present. Labeling data is often a manual and laborious task for a data scientist. This bares the danger of introducing a human bias on the model. Using semi-supervised methods can often increase the accuracy of the model and, at the same time, reduce the time and cost for producing a model since not all data needs to be labeled. For instance, manually labelling data for gene sequencing is an insurmountable task for a data scientist due to the large volumes of data; there are approximately six billion base pairs in a human, diploid genome. Semi-supervised methods can alleviate this problem by only labelling a small amount of data. In other words, the features are learned from a small amount of labelled data. First, we train a model with the small amount of labeled data. Then we label the unlabeled data set with the trained model that gave the best results. This yields a large volume of pseudo-labeled data. We now combine the labeled and pseudo-labeled data sets in order to train a model. This way, we should be able to better capture the underlying data distribution and the model generalizes better. As usual, there are other semi-supervised techniques. Semi-supervised methods give access to large amounts of data for training while minimizing the time spent for labeling training data. Ideally, semi-supervised methods can even reveal new labels.

Semi-supervised learning can be associated to human concept learning. Parents tell children the name of an object or animal when they see it the first time, they label it. The small amount of parental labelling suffices for children to be able to label new objects when they see them.

1.5.4 Machine learning and statistics

Machine learning uses a lot of statistics. That is why you will often find statistical terms and machine learning terms used interchangeably in literature. For instance, the terms dependent and independent variable are statistical terms, feature and label are their machine learning equivalents. Table 1.1 opposes some popular machine learning terms and their corresponding statistical terms.

The term statistical learning is sometimes confused with machine learning. There is a subtle difference between the two methods. Both

Table 1.1: Machine learning and statistical terminology.

Statistics	Machine Learning
Fitting	Learning
Classification, Regression	Supervised Learning
Clustering, Density estimation	Unsupervised Learning
Independent variable, covariate	Feature
Dependent variable, Response	Label
Observation	Instance

methods are data-based. However, machine learning uses data to improve the performance of a system. Statistical learning is based on a hypothesis. The basic assumption in statistical learning is that the data share some properties with a certain regularity. Therefore, statistical learning can use probabilistic methods and probabilistic distributions can be used to describe the statistical regularity. Machine learning is not based on assumptions. Statistical learning is math intensive and usually uses smaller data sets with fewer features, whereas machine learning can use billions of instances and features. Statistical learning requires a good understanding of the data, whereas machine learning can identify patterns iteratively with much less human effort. Therefore, machine learning is often preferred over statistical learning. However, since statistical learning requires us to familiarize ourselves with the data, it helps us gain more confidence in our model.

Machine learning is sometimes criticised as just being glorified statistics. Whereas it is true that machine learning uses a lot of statistics, machine learning also uses other concepts, such as data mining or neuro-computing, and is, as such, a subarea of computer science. Statistics is a subarea of mathematics. Both, machine learning and statistical learning are methods that learn from data. However, machine learning is an algorithm that learns from data without the need to be programmed by a developer. Statistical learning is a mathematical formalization of relationships between variables in the form of equations.

1.6   Data post-processing

Data post-processing includes data visualization (knowledge visualization), typically using a dashboard, reporting or dissemination of a result. A result is a useful or interesting pattern found during data mining. Data mining may generate thousands of patterns, not all of which are interesting. A pattern is interesting if it can easily be understood, if it is potentially useful, if it has been validated on new, unseen data or if it validates some hypothesis that you wanted to confirm. For instance, a model trained for predictive maintenance to detect a machine failure before it occurs should detect certain conditions, such as material fatigue or abrasion. It should be clear why a model predicts a machine failure and it should have been validated on test data.

It should be noted that these data mining steps are not comprehensive. Also, the steps do not need to be in the described order. Data cleansing can happen during data pre-processing, but it can also be necessary as a post-processing step in order to be able to interpret the result more easily or to increase readability before visualization. Also, supervised and unspuervised methods are not mutually exclusive. For instance, clustering can be used as preliminary step for a supervised learning task, for instance, to reduce the amount of data clustering can be used for data compression. It can also be used as a post-processing step, for instance, for data summarization.

Chapter 2

Machine Learning Basics

CONTENTS

Machine learning is one of many subfields of artificial intelligence or AI. Machine learning algorithms automatically build models that can identify useful patterns in data and make predictions without being explicitly programmed. Machine learning can be roughly divided into supervised, semi-supervised and unsupervised methods. Supervised methods are used for classification and regression, whereas unsupervised methods are used for clustering. Reinforcement learning is concerned with how agents act in an environment by maximizing some

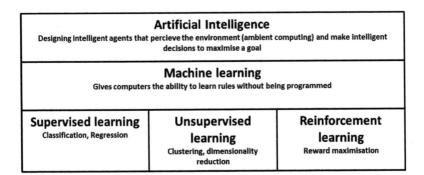

Figure 2.1: Machine learning areas.

notion of reward. Figure 2.1 shows the subareas of machine learning and summarizes the tasks. However, in literature, other classifications of machine learning methods can be found.

Machine learning tries to imitate human learning on a computer. Human learning is mostly based on experience, machine learning is based on data. Data is the starting point of a machine learning project and is used to train and test a model. For supervised methods, labelled data has to be present. Labeled data is data with correctly assigned tags. The labeled data is used for training and evaluating the accuracy of the learner. For instance, a learner that is supposed to do handwriting recognition needs a training data set with handwritten letters and their associated blockletters. The algorithm will then learn the relationship between the handwritten letters and the print letters and create a mapping between them. Unsupervised learners do not use labeled data, which makes evaluating a trained cluster algorithm more challenging.

Typical supervised methods include:

- Naïve Bayes
- Decision trees
- Neural networks (Multilayer perceptron)
- Linear regression
- Logistic regression
- Support vector machines
- k-nearest neighbor

Typical unsupervised methods include:

- k-means clustering
- Hierarchical clustering
- Principal component analysis (PCA)

The classification is not exclusive, since some learning schemes can be used for supervised and unsupervised tasks. For instance, artificial neural networks can be used in a supervised as well as in an unsupervised setting.

2.1   Supervised learning

Supervised learning needs labeled data for training. A training example is called a data point or instance and consists of an input and output pair (x, y). y is the output or ground truth for input x. Contrary to supervised learning, in unsupervised learning, the data only provides inputs. A multiset of training examples forms the training data set. The training data set is also called gold standard data and is as close to the ground truth as possible. The training set is used to produce a predictor function f, also called decision function, that maps inputs x to $y = f(x)$. The goal is to produce a predictor f that works with examples other than the training examples. In other words, f needs to generalize beyond the training data and provide accurate predictions on new, unseen data. f provides an approximation on unseen data. The input x is a feature vector or covariates, the output y is the response. A feature vector $\phi(x)$ is a map of feature names to feature values, i.e., strings to doubles. A feature vector $\phi(x) \in \mathbb{R}$ is a real vector $\phi(x) = [\phi_1(x), \phi_2(x), ... \phi_n(x)]$, where each component $\phi_i(x)$ with $i = 1, ... n$ represents a feature. The feature vector is computed by the feature extractor ϕ and can be thought of as a point in a high-dimensional feature space. A feature in a feature vector can be weighted, which means, not every feature necessarily contributes equally to a prediction. A weight is a real number that is multiplied with the feature value. The weights are represented in a separate weight vector because we want one single predictor that works on any input. Given a feature vector $\phi(x)$ and a weight vector w, we calculate the prediction score as the inner product of the vectors $w \cdot \phi(x)$. We do not know the weights in vector *w* beforehand, they are learned during training.

To predict, for instance, if a news article is about politics or sports, we need to find out what properties of input *x* might be relevant to predict if the article belongs to the category "politics" or "sports". This process is called feature extraction. There are machine learning techniques where no function is generated. For example, decision tree induction creates a decision tree, as the name suggests, usually represented as a dendrogram and not a function. To begin with, we will start with a function and explain other techniques in later chapters.

If we try to classify the data into categories, e.g., medical images into healthy or pathologic tissue, we have a classification problem. Classification has discrete outputs such as "true" or "false" or "0" or "1". If there are just two categories, it is a binary classification problem. Often, the outputs are probabilities, c.g., a mail is spam with 80 percent probability. In other words, the dichotomy of the output is not imperative.

If there are more categories, e.g., restaurant reviews using a Likert scale, as shown in Figure 2.2, it is a multiclass classification problem.

1. The food in the restaurant was excellent

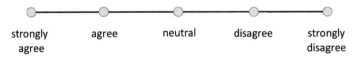

| strongly agree | agree | neutral | disagree | strongly disagree |

Figure 2.2: Likert scale.

If we have a binary classification problem, we are looking for a function *f* that will classify an input *x*, e.g., movie reviews, into "positive" and "negative" reviews, as represented by equation 2.1.

$$x \rightarrow f \rightarrow y \in \{+1, -1\} \tag{2.1}$$

where

x = Movie review

y = +1 ≡ positive; −1 ≡ negative

Technically, if the labels are +1 and −1 it could be considered a regression problem. If we have a multiclass classification problem, y is a category. For instance, predicting tomorrows weather is a multiclass classification problem where the categories could be "sunny", "rainy", "foggy", etc. If we want to predict how much it is going to rain tomorrow, we have a regression problem. Regression has continuous outputs.

Other prediction tasks include ranking and structured predictions. For instance, predicting who is going to win the marathon, who is going to be second and third, etc., as represented in equation 2.2, is a ranking problem and y is a permutation.

$$1,2,3,4 \rightarrow f \rightarrow 3,4,1,2 \qquad (2.2)$$

Structured predictions predict structured objects rather than real or discrete values. In structured predictions, the output is built from parts, such as a translation, where the output, the translation into another language, is built from words in the target language. Structured predictions have a wide variety of applications in natural language processing (NLP), speech recognition and computer vision, to name a few.

2.1.1 Perceptron

The perceptron is a simple neural network, proposed in 1958 by Rosenblatt, which became a landmark in early machine learning history [10]. It is one of the oldest machine learning algorithms. The perceptron consists of a single neuron and is a binary classifier that can linearly separate an input x, a real valued vector, into a single, binary output. It is an online learning algorithm. Online learning means it processes one input at a time. The data is presented in a sequential order and updates the predictor at each step using the best predictor. The perceptron is

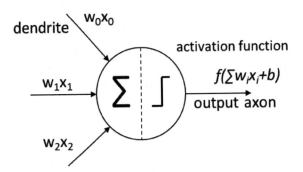

Figure 2.3: Perceptron.

modeled after a neuron in the brain and is the basic element of a neural network.

There are different types of perceptrons. They can have one or more input and output links. The input links correspond to the dendrites of a real neuron, the output links to the synapses. Also, there are different types of activation functions. The perceptron shown in Figure 2.3 consists of three weighted input links, a sum function, a step activation function and one output link.

A perceptron performs three processing steps:
1. Each input signal x is multiplied by a weight w: $w_i x_i$
2. All signals are summed up
3. The summed up inputs are fed to an activation function

The output is calculated using the output function $f(x)$, as defined by equation 2.3.

$$f(x) = \sum_{i=1} w_i x_i + b \qquad (2.3)$$

where
w = Weight
x = Input
b = Bias

A few remarks: Each input signal has its own weight that is adjusted individually. During training, the weights are adjusted until the desired result is obtained. Usually, the initial weights are all zero or assigned random numbers, usually between 0 and 1. After each learning iteration, the error at the output is calculated. The weights are adjusted based on the calculated error, in such a way that the next result is closer to the desired result.

The scaled up or down inputs are summed up to a single value. The summed up value is assigned an offset, called bias b. The perceptron also adjusts the bias during training. The bias shifts the activation curve to the left or right. However, the bias does not depend on any input.

The summed up input values are passed to an activation function. In it's most basic form, the activation function returns 1 if the input is positive, and 0, if it is negative, see equation 2.4. If 1 is returned, the perceptron is said to fire.

$$f(x) = \begin{cases} 1 & \text{if } wx + b > 0 \\ 0 & \text{otherwise} \end{cases} \tag{2.4}$$

The perceptron only returns 0 or 1. In other words, it is a binary classifier. It can only linearly separate a set of data points. A data point can be considered a point in a two dimensional plain. The input vector represents the coordinates of a point. The perceptron separates the data points into two sets using a line, the decision boundary, as shown in the left Cartesian coordinate system in Figure 2.4. We can separate, for instance, medical images of tissue into a set of images with normal and a set of images with pathologic tissue. However, not all binary classification problems are linearly separable. The points in the right coordinate system in Figure 2.4 cannot be separated linearly. Other machine learning algorithms, such as a multilayer perceptron or support vector machines, are capable of nonlinearly separating data points.

During training, the perceptron learns where to position the separation line. There can be an infinite number of possible decision boundaries, as shown in Figure 2.6 on the left. The perceptron cannot distinguish between different separation lines. A class of machine learning algo-

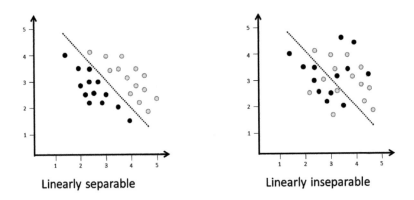

Figure 2.4: Linearly separable and inseparable data points.

rithms called margin classifiers also consider the distance between the data points and the decision boundary. Support vector machines fall into this category. This class of machine learning schemes tends to generalize better on new, unseen data.

A perceptron can have different types of activation functions. An activation function defines the output of the perceptron. Its simplest form is a step function, where the perceptron either fires or not. If the neural network is required to separate beyond linear separation, several perceptrons are required. An activation function can have a curved form, as shown in Figure 2.5 on the right side. This type of activation function is called a sigmoid activation function and is capable of firing with different rates. As the input increases, it fires at an increasing rate. We will see other types of activation functions such as rectifiers often used for deep learners in later chapters.

Not all data points in Figure 2.6 are correctly separated by the decision boundary. Some points are on the wrong side of the line. This reflects the fact that machine learning algorithms often return an approximation, not a perfect separation. A classifier that perfectly separates the training data points often does not work well on new data. A learner needs to generalize in order to work on data other than the training data. Otherwise, we have an overfit learner.

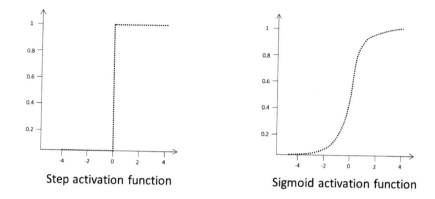

Figure 2.5: Activation functions.

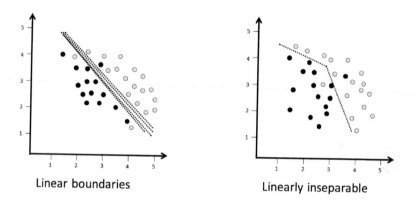

Figure 2.6: Linear and nonlinear fitting.

The planes in Figure 2.6 have just two dimensions. Coordinate systems with more than two dimensions are needed when there is more than one feature. This depends on the input vector and on the learner that is being used. For better presentability, we will often use just two dimensions, even if more are possible or needed.

2.2 Unsupervised learning

In supervised learning, we have input and output data. In unsupervised learning, we do not have labeled training data. In other words, we only have input data, no output data, i.e., there is no y. Instead of predicting or approximating a variable, we try to group the data into different classes. For instance, we have a population of online shop users that we try to segment into different types of buyers. The types could be social shopper, lifestyle junkie and detached introvert. In this case, we have three clusters. If an observation point can only belong to one cluster, it is an exclusive cluster, if it is allowed to be part of more than one cluster, it is a non-exclusive cluster, in other words, the clusters can be overlapping.

Let N be a set of unlabeled instances $D = x_1, x_2, ..., x_N$ in a d-dimensional feature space, in clustering D is partitioned into a number of disjoint subsets D_js:

$$D = \cup_{j=1}^{k} D_j \tag{2.5}$$

where

$$D_i \cap D_j = \emptyset$$
$$i \neq j$$

The points in each subset D_j are similar to each other according to a given criterion $\emptyset$. Similarity is usually expressed by some distance measure, such as the Euclidean distance or Manhattan distance.

Cluster analysis can be used for understanding the underlying structure of a dataset by grouping it into classes based on similarities. Grouping similar objects into conceptually meaningful classes has many real-life applications. For instance, biologists spent a considerable amount of time on grouping plants and animals into classes, such as rodents, canines, felines, ruminants, marsupials, etc. Clustering is also used for information retrieval, for instance, by grouping Web search results into categories, such as news, social media, blogs, forums, marketing, etc. Each category can then be divided into subcategories. For example, news is divided into politics, economics, sports, etc. This process is called hierarchical clustering. Humans, even young children, are skilled at grouping objects into their corresponding classes. This capability is important to humans for understanding and describing the world around us. This is crucial for our survival. We need to know which groups of animals are harmful or which traffic situations are potentially dangerous.

Cluster analysis can also be used as an intermediate step for other data analysis tasks, such as outlier removal, data summarization, data compression or as input for other machine learning algorithms, such as the k-nearest neighbor algorithm. In data summarization, instead of applying a machine learning algorithm to the entire data set, it can be applied to each cluster individually, thus, reducing processing power and time. This can speed up computing time for machine learning schemes, such as k-nearest neighbor. Outliers can have a negative impact on some algorithms, e.g., regression models. Clustering can be used as a preliminary step for outlier removal in regression analysis.

2.2.1 k-means clustering

k-means clustering is one of the most popular cluster and machine learning algorithm. k-means clustering falls into the category of centroid-based algorithms. A centroid is the geometric center of a geometric plane figure. The centroid is also called barycenter. In centroid-based clustering, n observations are grouped into k clusters in such a way that each observation belongs to the cluster with the nearest centroid. Here, the criterion for clustering is distance. The centroid itself does not need to be an observation point. Figure 2.7 shows k-means clustering with 3 clusters.

In k-means clustering, the number of clusters k needs to be defined beforehand, which is sometimes a problem since the number of clusters might not be known beforehand. For instance, in biology, if we want to classify plants given their features, we might not know how many different types of plants there are in a given data set. Other clustering methods, such as hierarchical clustering, do not need an assumption on the number of clusters. Also, there is no absolute criterion; it has to be defined by the user in such a way that the result of the clustering will suite the aim of the analytics task at hand. Cluster analysis is often just the starting point for other purposes, for instance a pre-processing step for other algorithms.

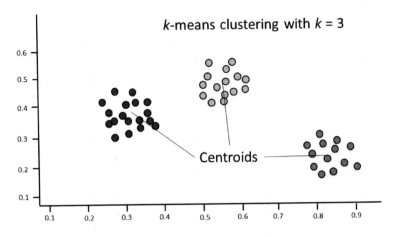

Figure 2.7: k-means clustering.

Humans and animals learn mostly from experience, not from labeled data. Humans discover the world by observing it, not by being told the label of every object. So, human learning is largely unsupervised. That is why some authors argue that unsupervised learning will become fare more important than supervised learning in the future [18].

2.3 Semi-supervised learning

Semi-supervised learning uses a combination of labeled and unlabeled data. It is typically used when a small amount of labeled data and a large amount of unlabeled data is present. Since semi-supervised learning still requires labeled data, it can be considered a subset of supervised learning. However, other forms of partial supervision other than classification are possible.

Data is often labeled by a data scientist, which is a laborious task and bares the risk of introducing a human bias. Bias stems from human prejudice that, in the case of labeling data, might result in the under- or overrepresentation of some features in the data set. Machine learning heavily depends on the quality of the data. If the data set is biased, the result of the machine learning task will be biased. However, human bias cannot only negatively influence machine learning through data but also through algorithms and interaction. Human bias does not need to be conscious. It can originate from ignorance, for instance, an under-representation of minorities in a sample population or from a lack of data that includes minorities. Generally speaking, training data should equally represent all our world.

There are many different semi-supervised training methods. Proba-bly the earliest idea about using unlabeled data in classification is self-learning, which is also known as self-training, self-labeling, or decision-directed learning [4]. The basic idea of self-learning is to use the labeled data set to train a learning algorithm, then use the trained learner iteratively to label chunks of the unlabeled data until all data is labeled using pseudo labels. Then the trained learner is retrained us-ing its own predictions. Self-learning bears the risk that the pseudo-labeled data will have no effect on the learner or that self-labeling happens without knowing on which assumptions the self-learning is based on. The central question for using semi-supervised learning is, under which conditions does taking into consideration unlabeled data contribute to the prediction accuracy? In the worst case, the unlabeled data will deteriorate the prediction accuracy.

2.4 Function approximation

Function approximation (FA) is sometimes used interchangeably with regression. Regression is a way to approximate a given data set. Func-tion approximation can be considered a more general concept since there are many different methods to approximate data or functions. The goal of function approximation is to find a function f that maps an input to an output vector. The function is selected among a defined class, e.g., quadratic functions or polynomials. For instance, equation 4.14 is a first degree polynomial equation. Contrary to function fitting, where a curve is fitted to a set of data points, function approximation aims to find a function f that approximates a target function. Given one set of samples (x, y), we try to find a function f:

$$f : X \rightarrow Y \tag{2.6}$$

where

X	$=$	Input space
Y	$=$	Output space, number of predictions

where the input space can be multidimensional $\mathbf{X} \subseteq \mathbb{R}^2$, in this case n-dimensional. The function:

$$f(x) = y \qquad (2.7)$$

maps $x \in X$ to $y \in Y$, where the distribution of x and the function f are unknown. f can have some unknown properties for a space where no data points are available.

Figure 2.8 shows an unknown function $f(x)$ and some random data points.

Function approximation is similar to regression and techniques such as interpolation, extrapolation or regression analysis can be used. Regression does essentially the same thing, create a model from a given data set. However, regression focuses more on statistical concepts, such as variance and expectation. Function approximation tries to explain the underlying data by finding a model $h(x)$ for all samples (x, y), such that $h(x) \approx y$. Ideally, the model equals the underlying function $f(x)$ such

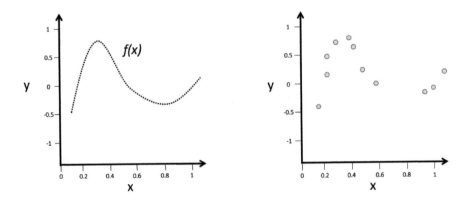

Figure 2.8: Function approximation.

that $h(x) = f(x)$. However, there might be sub-spaces where no data points are available. To find a model, we need to measure the quality of the model. To do so, function approximation tries to minimize the prediction error. This is similar to evaluating a trained machine learning model, such as a Bayesian or a logistic regression model. An intuitive measure for the error is the mean absolute error (MAE). Given n instances (x, y), the model $h(x)$ can be evaluated by calculating the mean absolute error as:

$$MAE = \frac{1}{n} \sum_{i=1}^{n} |y_i - h(x_i)| \qquad (2.8)$$

which measures the average deviation from the actual value. However, the mean absolute error does not take variance into account. It treats all errors evenly according to their magnitude. The mean absolute error is one of a number of ways to compare the predicted value with their eventual outcome. Other well established alternatives include the mean squared error or the root mean squared error. It is not always easy to decide which measure is appropriate. The squared error measures and root squared error measures weigh large discrepancies much more heavily than small ones, whereas the absolute error measures do not [40]. Fortunately, in practical situations, the best prediction model is still the best, no matter which error measure is used.

In the simplest form of function approximation, the type of function is known. When the underlying function type is known, e.g., polynomial, exponential or elliptic, only the parameters of the function have to be found and we have a curve fitting or model fitting problem. For instance, if the function $y = ax^2 + bx + c$ has to be fitted to a given set of data points, we have to find the coefficients a, b and c. However, the underlying function type is often not known and needs to be found. The chosen function type is a best guess and a wrong assumption about the function type might result in poor quality models. Instead of using a predefined function, arbitrary models can be built using non-parametric regression. Using approaches such as interpolation, the function does not need to be specified in advance. The draw back is that learning takes considerably longer. There are many function approximation approaches and describing them is beyond the scope of this book. However, some of the machine learning approaches, such as artificial neural networks (ANN) and regression, can be used for function approximation and will be described in later chapters.

2.5 Generative and discriminative models

Predictive algorithms are often categorized into generative and discriminative models. This distinction presumes a probabilistic perspective on machine learning algorithms. Generally speaking, a generative model learns the joint probability distribution $p(x,y)$, whereas a discriminative model learns the conditional probability distribution $p(y|x)$, in other words, the probability of y given x. A generative algorithm models how the data was generated in order to categorize a signal. It is called generative since sampling can generate synthetic data points. Generative models ask the question: Based on my generation assumptions, which category is most likely to generate this signal? Generative models include naïve Bayes, Bayesian networks, Hidden Markov models (HMM) and Markov random fields (MRF).

A discriminative algorithm does not care about how the data was generated, it simply categorizes a given signal. It directly estimates posterior probabilities, they do not attempt to model the underlying probability distributions. Logistic regression, support vector machines (SVM), traditional neural networks and nearest neighbor models fall into this category. Discriminative models tend to perform better since they solve the problem directly, whereas generative models require an intermediate step. However, generative models tend to converge a lot faster.

2.6 Evaluation of learner

In order to evaluate the quality of a trained model, we need to evaluate how well the predictions match the observed data. In other words, we need to measure the quality of the fit. The quality is measured using a loss function or objective function. The goal of all loss functions is to measure how well an algorithm is doing against a given data set.

The loss function quantifies the extend to which the model fits the data [14], in other words, it measures the goodness of fit. During training, the machine learning algorithm tries to minimize the loss function, a process called loss minimization. We are minimizing the training loss, or training error, which is the average loss over all the training samples. A loss function calculates the price, the loss, paid for inaccuracies in a classification problem. It is, thus, also called the cost function.

There are many different loss functions. We have already seen the mean absolute error in equation 2.8. Another popular loss function is the mean squared error (MSE) since it is easy to understand and implement. The mean squared error simply takes the difference between the predictions and the ground truth, squares it, and averages it out across the whole data set. The mean squared error is always positive and the closer to zero the better. The mean squared error is computed as shown in equation 2.9.

$$MSE = \frac{1}{n} \sum_{i=1}^{n} (y_i - f(x_i))^2 \qquad (2.9)$$

where

y	=	Vector of n predictions
n	=	Number of predictions
$f(x_i)$	=	Prediction for the ith observation

Other popular loss functions include hinge loss or the likelihood loss. The likelihood loss function is also a relatively simple function and is often used for classification problems. It just multiplies the predicted probabilities for each instance. It is not as easily interpreted as the mean squared error, but it is useful for comparing models. As mentioned before, it is not always obvious which loss function to use. It depends on the problem and the data set at hand. For instance, the squared loss penalizes large residuals a lot more than other types of loss functions. Also, it should be noted that other evaluation methods apply for learners such as decision trees. For decision trees, we need a measure for the purity, which means, how well the classes are separated. The Gini impurity or the information gain entropy can be used.

Bias and variance

The bias-variance tradeoff problem is a well known problem in supervised learning. It can be used to understand under- and over-fitting. Models with high bias tend to be less complex but underfit the data and miss important information in the data. Bias is an error that stems from over simplistic assumptions. Models with high variance tend to be more sensitive and overfit the data. They represent the training data well but do not generalize well beyond the training data. They tend to capture noise instead of the signal. Ideally, both bias and variance should tend towards zero. The bias-variance trade-off refers to the fact that it is typically impossible to minimize both at the same time. Bias and variance will be explained in more detail in Chapter 9.1.1.

2.6.1 Stochastic gradient descent

During training we try to minimize a loss function. So we need to make sure we learn into the right "direction" and not getting worse. For a simple loss function, we could use calculus to find the optimal parameters that minimize the loss function. However, for a more complex loss function, calculus might not be feasible anymore. One approach is to iteratively optimize the objective function by using the gradient of the function. This iterative optimization process is called gradient descent where the gradient tells us which direction to move to decrease the loss function. Figure 2.9 shows gradient descent with a hypothetical objective function w^2 and the derivative $2w$ that gives us the slope. If the derivative is positive, it slopes downwards to the left as shown in Figure 2.9, if it is negative, it slopes to the right. The value of the derivative is multiplied by the constant step size η, called the learning rate, and subtracted from the current value. Once the change of the parameter value becomes too small, in Figure 2.9 when it approaches zero, the process stops. The hyperparameter η is the step size, which defines how fast we move to the minimum.

The problem with gradient descent is that it is slow since every iteration has to go through all the training instances, the batch, which is expensive, especially if the training set is large. The training loss is the

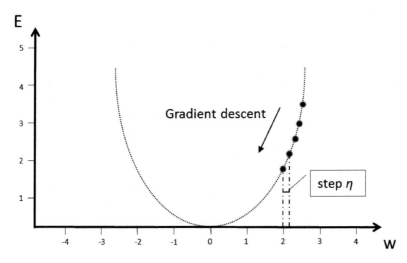

Figure 2.9: Gradient descent.

sum over all the training data, which means, the algorithm has to go through all training instances to move one step down.

A much more efficient method is stochastic gradient descent (SGD). Stochastic gradient descent addresses the issue of slow convergence when the training data set is large and when it contains a large number of features. Furthermore, in an online setting, there is no easy way to add data in batch methods, where the entire training set is used for each step. In Stochastic gradient descent, instead of using the entire batch in every iteration, the gradient is estimated using only one instance, a batch of size 1, for every step. In other words, we use the cost gradient of one instance in each iteration, instead of using the sum of the cost gradient of the full batch. The term stochastic refers to the fact that the single instance that is used per iteration is chosen randomly. The assumption is that, given enough iterations, using only one instance per iteration, we can estimate the gradient on average. Each iteration might not be as good as an iteration using the full batch, but this way we can make many iterations and in practice, in many situations making just one pass over each training example often performs equally well to going through, say, ten iterations over the full batch. Some implementations use more than just one training instance. Mini-batches of ten to one thousand instances are a compromise between using just one instance and a full batch. Since very large batches are often noisy, stochastic gradient descent using single or mini-batches still converges fast and can reduce variance. Some implementations adapt the step size for every iteration.

A few remarks: Choosing the step size η is crucial since it defines how fast the algorithm converges. Using a large η the algorithm might converge faster than with a small η. However, with η set too large, there is the risk that the search overshoots and the algorithm never reaches the minimum. If the step size is too small and the error function has several minima (typical for loss functions of artificial neural networks) as shown in Figure 2.10, it might get caught in a local minimum and never reach the absolute minimum. To avoid this, some implementations adapt the step size as mentioned before. A suggested approach sets the step size to one and decreases it during training. In other approaches, the learning rate is halved as convergence slows down or it can be annealed at each iteration. Many extensions and variations have been proposed, since setting the learning rate is crucial to obtain good results.

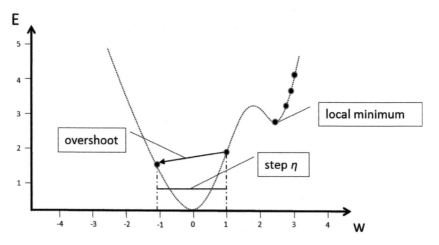

Figure 2.10: Gradient descent issues.

In stochastic gradient descent, the path to the minimum is noisier, i.e., more random, than when using the full batch. However, we are indifferent to the path as long as we get to the minimum faster. Using the full batch does a lot of redundant calculations on large data sets since it calculates the gradients of similar examples. When using one instance in each step, the redundancy goes away. That is why stochastic gradient descent usually converges faster from gradient descent.

Stochastic gradient descent is the most common optimization method used in machine learning. It is also used for deep learners, which have been making a lot of progress in the past years and which are, in essence, artificial neural network's with many hidden layers. Slightly modified versions of the stochastic gradient descent algorithm remain the dominant training algorithms for deep learning models today [8].

2.6.2   *Cluster evaluation*

The goal of clustering is to partition data in a way to attain a high intra-cluster similarity. The ideal cluster is dense and isolated. Since we do not have labeled data in clustering, we cannot use the same optimization methods used in supervised learning. Evaluating a cluster is not as straight forward as evaluating a classifier since there is no ground truth. Since there are usually several different possible ways to cluster a data set, depending on the similarity metric chosen, human judgment is required in order to evaluate the quality of a cluster. There are internal measures, such as the Dunn index, that measure the cluster densities and the well separateness of a cluster. The Dunn index measures the ratio between the minimal and maximal inter-cluster distance for each cluster partition. Other internal evaluation methods that use some form of distance measure are the Silhouette coefficient and the Davies-Bouldin index. However, since distance can be used as similarity measure, using a distance metric for cluster evaluation might overrate the resulting cluster.

External methods use data that was not used for clustering as benchmark. These methods use benchmark data to measure how close the cluster is to the benchmark. The benchmark data is pre-classified, often by a data scientist, and can be considered the gold standard for clustering. However, pre-classified clusters require "knowledge" that is not present and that we might want to discover during clustering. As with internal methods, there are several methods using external measures. They include purity, the Rand index and the F-measure. Purity measures the extend to which a cluster only contains data points belonging to a single class. It is simple and transparent. To calculate purity, each cluster is assigned to the class most frequent in the cluster. Then, for each cluster w, the correctly assigned data points to class c are counted. For a given set of clusters $W = w_1, w_2, ..., w_k$ and classes $C = c_1, c_2, ..., c_j$ partitioning N data points, the purity can be defined as

$$purity(W,C) = \frac{1}{N} \sum_k \max_j |w_k \cap c_j| \qquad (2.10)$$

Good clusters have a purity close to 1, bad clusters close to 0. If the number of clusters is large, purity is high since purity does not penalize a high number of clusters. If every data point is separated out in it's own cluster, purity is 1. The Rand index measures the percentage of correctly assigned data points, in other words, it measures accuracy. It penalizes both false positive (FP) and false negative (FN) decisions. In addition to this, the F-measure balances the contribution of false negative through a parameter by penalizing false negatives more than false positives. It supports differential weighting of these two types of errors.

As mentioned before, evaluating a cluster is not an easy task and it is not always obvious which method works best. A good internal measure does not necessarily translate into good performance in a real world application. A more effective approach is evaluating a cluster directly in the application. For instance, if we cluster search results we can measure the time it took to find an answer. However, this might be too time consuming and expensive. So, we can use an external method as a substitute for human judgment. If we search for "Apple" and have two clusters with search results, one for apple the fruit, and one for Apple Inc. the computer company, we only use the cluster provided by the gold standard and not the class label. Since clustering has many applications in medicine, market research, web search, recommendation engines, etc., just to name a few, which evaluation method to select needs to be decided based on the intended use of the project and on the data set to be analyzed. Clustering is not an automatic task but an iterative optimization task that relies on trial and error.

MACHINE
LEARNING

Chapter 3

Data Pre-processing

CONTENTS

Real world data is seldom in a form suitable for machine learning algorithms. It is typically noisy, contains redundant and irrelevant data, or the data set is too large to be processed efficiently in its original form. Data pre-processing extracts the relevant information, the features, from a data set and transforms it into a form that can be used as input for an machine learning algorithm, usually in the form of a feature vector. Data pre-processing reduces the cost of feature measurement, increase the efficiency of a learner, and allows for higher classification accuracy.

Machine learning algorithms are domain independent, data pre-processing is highly domain specific and usually requires domain knowledge. Data pre-processing comprises a large number of different techniques and often several training data sets are created by using different pre-processing techniques. The learner is then trained with each of the pre-processed training sets and the best performing set is used for the final model. Some tools offer feature filtering so the same data set can be tested with different features.

Typical pre-processing tasks include data deduplication, outlier removal, stop word removal from texts, where irrelevant words or signs are removed from the original text. Sometimes data is enriched with synthetic data or synthetic features, especially for sparse data sets.

3.1   Feature extraction

Feature extraction, also called feature selection or feature engineering, is one of the most important tasks to find hidden knowledge or business insides. In machine learning, rarely is the whole initial data set that was collected used as input for a learning algorithm. Instead, the initial data set is reduced to a subset of data that is expected to be useful and relevant for the subsequent machine learning tasks. Feature extraction is the process of selecting values from the initial data set. Features are distinctive properties of the input data and are representative descriptive attributes. In literature, there is often a distinction between feature selection and feature extraction [43]. Feature selection means reducing the feature set into a smaller feature set or into smaller feature sets, because not all features are useful for a specific task. Feature extraction means converting the original feature set into a new feature set that can perform a data mining task better or faster. Here, we treat feature extraction and feature selection interchangeably. Feature selection and, as mentioned before, generally data pre-processing is highly domain specific, whereas machine learning algorithms are not. Feature selection is also independent of the machine learning algorithm used.

Features do not need to be directly observable. For instance, a large set of directly observable features might be reduced using dimensionality reduction techniques into a smaller set of indirectly observable features, also called latent variables or hidden variables. Feature extraction can be seen as a form of dimensionality reduction. Often features are weighted, so that not every feature contributes equally to the result. Usually the weights are presented to the learner in a separate vector.

A sample feature extraction task is word frequency counting. For instance, reviews of a consumer product contain certain words more often if they are positive or negative. A positive review of a new car typically contains words such as "good", "great", "excellent" more often than negative reviews. Here, feature extraction means defining the words to count and counting the number of times a positive or negative word is used in the review. The resulting feature vector is a list of words with their frequencies. A feature vector is an n-dimensional vector of numerical features, so the actual word is not part of the feature vector itself. The feature vector can contain the hash of the word as shown in Figure 3.1, or the word is recognized by its index, the position in the vector.

Another example of feature extraction is object recognition in an image. The region of the object of interest has to be detected in the image and the shape has to be isolated. The background of the region is omitted. The relevant region is then converted into a pixel array which can be used as the input vector for a learning scheme.

As with machine learning schemes, there are many different feature extraction techniques and there are no simple recipes. A lot of research has been conducted in automatic feature extraction, also called feature

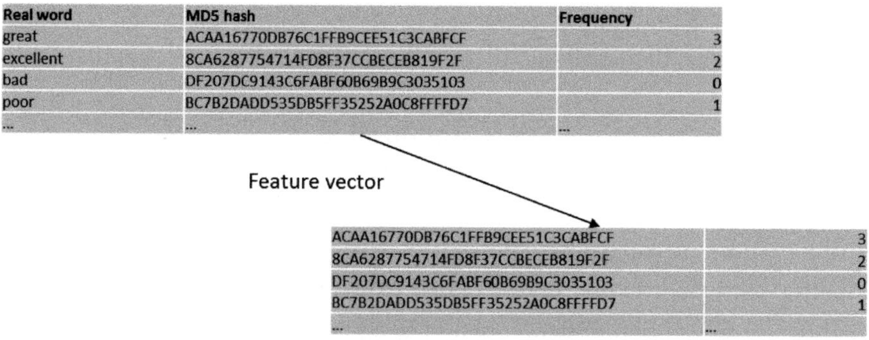

Real word	MD5 hash	Frequency
great	ACAA16770DB76C1FFB9CEE51C3CABFCF	3
excellent	8CA6287754714FD8F37CCBECEB819F2F	2
bad	DF207DC9143C6FABF60B69B9C3035103	0
poor	BC7B2DADD535DB5FF35252A0C8FFFFD7	1
...	...	...

Feature vector

ACAA16770DB76C1FFB9CEE51C3CABFCF	3
8CA6287754714FD8F37CCBECEB819F2F	2
DF207DC9143C6FABF60B69B9C3035103	0
BC7B2DADD535DB5FF35252A0C8FFFFD7	1
...	...

Figure 3.1: Feature vector.

learning. Since feature extraction is a laborious and time consuming task, automatic feature selection can potentially speed up a data mining project considerably. In recent years, great advances have been made using deep learners. Deep learners have the capability to automatically extract features. Deep learners have learned how to play games such as chess without being trained to learn the rules first. They learned the rules through trial and error. Also, deep learners such as convolutional neural networks can extract the relevant areas in an image by themselves without any image pre-processing. Next to the astoundingly good results of deep learners in tasks such as multimedia mining and natural language processing, automatic feature extraction is one of the most exciting advancements in machine learning.

3.2 Sampling

Data sampling is the process of selecting a data subset, since analyzing the whole data set is often too expensive. Sampling reduces the number of instances to a subset that is processed instead of the whole data set. Samples can be selected randomly so every instance has the same probability to be selected. The selection process should guarantee that the sample is representative of the distribution that governs the data, thereby ensuring that results obtained on the sample are close to ones obtained on the whole data set [43].

There are different ways data can be sampled. When using random sampling, each instance has the same probability to be selected. Using random sampling, an instance can be selected several times. If the randomly selected instance is removed from the data set, we end up with no duplicates in the sample.

Another sampling method uses stratification. Stratification is used to avoid overrepresentation of a class label in the sample and provide a sample with an even distribution of labels. For instance, a sample might contain an overrepresentation of male or female instances. Stratification first divides the data in homogeneous subgroups before sampling. The subgroups should be mutually exclusive. Instead of using the whole population, it is divided into subpopulations (stratum), a bin with female and a bin with male instances. Then, an even number of samples is selected randomly from each bin. We end up with a stratified random sample with an even distribution of female and male instances. Stratification is also often used when dividing the data set into a training and testing subset. Both data sets should have an even distribution of the features, otherwise the evaluation of the trained learner might give flawed results.

3.3   Data transformation

Raw data often needs to be transformed. There are many reasons why data needs to be transformed. For instance, numbers might be represented as strings in a raw data set and need to be transformed into integers or doubles to be included into a feature vector. Another reason might be the wrong unit. For instance, Fahrenheit needs to be converted into Celsius, or inch needs to be converted into metric.

Sometimes the data structure has to be transformed. Data in JSON (JavaScript Object Notation) format might have to be transformed into XML (Extensible Markup Language) format. In this case, data mapping has to be transformed since the metadata might be different in the JSON and XML file. For instance, in the source JSON file names might be denoted by "first name", "last name" whereas in the XML it is called "given name" and "family name". Data transformation is almost always needed in one or another form, especially if the data has more than one source.

3.4   Outlier removal

Outlier removal is another common data pre-processing task. An outlier is an observation point that is considerably different from the other instances. Some machine learning techniques, such as logistic regression, are sensitive to outliers, i.e., outliers might seriously distort the result. For instance, if we want to know the average number of Facebook friends of Facebook users we might want to remove prominent people such as politicians or movie stars from the data set since they typically have many more friends than most other individuals. However, if they should be removed or not depends on the aim of the application, since outliers can also contain useful information.

Outliers can also appear in a data set by chance or through a measurement error. In this case, outliers are a data quality problem like noise. However, in a large data set outliers are to be expected and if the number is small, they are usually not a real problem. Clustering is often used for outlier removal. Outliers can also be detected and removed visually, for instance, through a scatter plot, or mathematically, for instance, by determining the z-score, the standard deviations by which the outlier is above the mean value of the data set.

3.5 Data deduplication

Duplicates are instances with the exact same features. Most machine learning tools will produce different results if some of the instances in the data files are duplicated, because repetition gives them more influence on the result [40]. For example, Retweets are Tweets posted by a user that is not the author of the original Tweet and have the exact same content as the original Tweet except for metadata such as the timestamp of when it has been posted and the user who posted, retweeted, it. As with outliers, if duplicates should be removed or not depends on the context of the application. Duplicates are usually easily detectable by simple comparison of the instances, especially if the values are numeric, and machine learning frameworks often offer data deduplication functionality out of the box. We can also use clustering for data deduplication since many clustering techniques use similarity metrics and they can be used for instance matching based on similarities.

3.6 Relevance filtering

Relevance filtering typically happens at different stages of a machine learning project. Data deduplication can be considered a relevance filtering step if every instance has to be unique. Feature selection can also be considered relevance filtering since relevant features are separated from irrelevant ones. Stop words removal in text analysis is a relevance filtering procedure since irrelevant words or signs such as smileys are removed. Many natural language processing frameworks offer stop words removal functionality. Stop words are usually the most common words in a language such as "the", "a", or "that". However, the list often needs to be adjusted since a stop word might be relevant, for instance, in a name such as "The Beatles".

Since feature selection can be considered a search problem, using different search filters can be used to combat noise. For instance, people often enter fake details when entering personal data, such as fake addresses or phone numbers, since they do not want to be contacted by a call center. These fake profiles need to be filtered out otherwise they can negatively influence the predictive performance of a learner. Often this already happens when data is collected by using queries that omit irrelevant or fake data.

Relevance filtering can also happen after the features have been selected. Different features often do not contribute equally to the result. Some features might not contribute at all and can be filtered out. Data mining tools usually provide filter functionality at the feature level so learners can be trained on different feature sets.

3.7   Normalization, discretization and aggregation

Normalization can mean different things in statistics. It can mean transforming data, that has been measured at different scales into a common scale. Using machine learning algorithms, numeric features are often scaled into a range from 0 to 1. Normalization can also include averaging of values, e.g., calculating the means of a time series of data over specific time periods, such as hourly or daily means. Sometimes, the whole probability distribution is aligned as part of the normalization process.

Discretization means transferring continuous values into discrete values. The process of converting continuous features to discrete ones and deciding the continuous range that is being assigned to a discrete value is called discretization [43]. For instance, sensor values in a smart building in an Internet of Things (IoT) setting, such as temperature or humidity sensors, are delivering continuous measurements, whereas only values every minute might be of interest. An other example is the age of online shoppers, which are continuous and can be discretized into age groups such as "young shoppers", "adult shoppers" and "senior shoppers".

Data aggregation means combining several feature values in one. For instance, going back to our Internet of Things example, a single temperature measurement might not be relevant but the combined temperature values of all temperature sensors in a room might be more useful to get the full picture of the state of a room.

Data aggregation is a very common pre-processing task. Among the many reasons to aggregate data are the lack of computing power to process all values, to reduce variance and noise and to diminish distortion.

3.8   Entity resolution

Entity resolution, also called record linkage, is a fundamental problem in data mining and is central for data integration and data cleaning. Entity resolution is the problem of identifying records that refer to the same real-world entity and can be an extremely difficult process for computer algorithms alone [39]. For instance, in a social media analysis project, we might want to analyse posts of users on different sites. The same user might have the user name "John" on Facebook, "JSmith" on Twitter and "JohnSmith" on Instagram. Here, entity resolution aims to identify the user accounts of the same user across different data sources, which is impossible if only the user names are known. Also, there is the danger that users are confused and the user name "JSmith" is associated with a different user, e.g., "James Smith". In this case, record disambiguation methods have to be applied. If the data set is large and we have n records, every record has to be compared with all the other records. In the worst case, we have $O(n^2)$ comparisons to compute. We can reduce the amount of comparisons by applying more intelligent comparison rules. For instance, if we have three instances a, b and c, if $a = b$ and $a \neq c$ we can infer that $b \neq c$. Reducing the number of comparisons can diminish the effort but is not always feasible and a considerable amount of research has been conducted to develop automated, machine-based techniques.

As with many pre-processing tasks, we can use clustering methods for entity resolution. In fact, entity resolution is a clustering problem since we group records according to the entity they belong to. It can be addressed similar to data deduplication by finding some similarity measures and then using a distance measure, such as the Eucledian distance or the Jaccard similarity, to find records that belong to the same real-world entity. Clustering techniques are described in more detail in Chapter 6. In practice, the probability that a record belongs to a certain entity is usually calculated. Entity resolution can also be used for reducing redundancies in data sets and reference matching, where noisy records are linked to clean ones. Active learning methods and semi-supervised techniques have also been used for entity resolution. However, machine-based techniques, despite all the research effort that has been invested, are far from being perfect.

Chapter 4

Supervised Learning

CONTENTS

In this chapter, supervised learning methods to make predictions are considered. "Supervised" refers to the fact that labeled training data is required. For instance, for classifying news articles into different categories such as "Politics", "Business", "Sports", etc., the learner is trained with a set of pre-classified, labeled, news articles. The categories are called "labels". During training, the learner creates an internal representation that represents a generalization of the training data. The process of creating internal representations is called induction. If no labeled training data is present, unsupervised learning applies. When using unsupervised methods, the learner must discover the categories during training. If a small amount of labeled data and a large amount of unlabeled data is present, semi-supervised learning methods can be used. Typical supervised learning algorithms include artificial neural networks, support vector machines, k-nearest neighbor (k-NN), Bayesian models or decision tree induction. Typical unsupervised learning algorithms include hierarchical clustering and k-means clustering. Since there is no labeled data in unsupervised learning, the accuracy of the trained learner cannot be evaluated. Measures such as density estimation or well separateness are used instead. The majority of machine learning projects use supervised methods.

In machine learning, an algorithm is a repeatable process to obtain a trained model given a set of data. Each algorithm behaves very differently and can be effective for different problems. One of the key distinctions of supervised learning algorithms is how much bias and variance they produce. These are called prediction errors. Bias is the difference between the expected and the actual, measured prediction. Variance refers to an algorithms sensitivity to a particular set of training data. During training, both errors, bias and variance, are reduced until they converge. A third prediction error, the irreducible error, stems from noise in the data or randomness. It cannot be reduced during training. However, better data cleansing steps might diminish the irreducible error in some cases.

Supervised learning is used for classification and regression. Classification is used when the target features are discrete, regression is used when the target features are continuous. For instance, predicting tomorrows weather is a classification task. The classes could be "sunny", "rainy", "snowy", etc. Predicting how many millimeters it will rain tomorrow is a regression task. In this example, the amount of rain is the dependent variable, whereas the input variables, e.g., barometric pressure, are the independent variables or predictors. Typically, many supervised learning schemes (but not all) calculate probabilities, e.g., the probability that it is going to rain tomorrow.

The majority of analytic problems are addressed using classification techniques, since they can be applied to support decision making. For instance, should we contact a specific customer in a marketing campaign or should we invest in a certain startup are typical classification problems. Some machine learning techniques, such as k-nearest neighbor, can be used for both classification and regression.

Supervised learning requires, next to training data:
∎ Input features
∎ Target features
∎ Measure of improvement

The input features, typically represented as an input vector, are extracted from the training data. The target feature(s) is the output or response variable(s). The measure of improvement is used to determine if we are getting better in every learning cycle. During training, the machine learning scheme learns to map the input features to target features. In other words, the internal representation of the learner is adjusted until the desired accuracy is reached. Often, the mapping is an approximation and 100% accuracy is not achieved. Accuracy is a performance metric for a learner. Many other evaluators exist, such as precision and recall or the receiver operating characteristic. Some will be discussed in section 5.1.

A full training cycle typically goes through the following steps:
1. Data collection
2. Data pre-processing
3. Training a machine learning algorithm, i.e., fitting model parameters
4. Testing/evaluating trained learning scheme
5. Measuring error

Training a machine learning algorithm is an iterative process and typically goes through many iterations until satisfactory results are obtained. Also, often several machine learning algorithms are trained and evaluated. The best performing model is then used to make predictions on new, unseen data. Usually, it is good practice to start with a simple model, such as a Bayesian model or the k-nearest neighbor learner, that can then be used as benchmark. If the results are not satisfactory, more sophisticated learning schemes can be trained and evaluated. If two learners perform equally well, the simpler one should be chosen in accordance with the Occam's razor principle. The Occam's razor principle states that if two problem solving solutions exist, the simpler one should be chosen. The ultimate goal is to have a model that isolates the signal and ignores the noise. How closely a models predicted values match the observed values is referred to as goodness of fit.

4.1 Classification

In classification, we try to assign a label to a test instance, e.g., we try to predict if the animal in a picture is a "cat" or a "dog". In other words, we assign a new observation to a specific category. The learning algorithm that assigns the instance to a category is called classifier. Classification answers questions such as: "is that bank client going to repay the loan?", "will the user who clicked on an ad buy?", "who is the person in the Facebook picture?". Classification predicts a discrete target label y. If there are two labels such as "spam" / "not spam", we have a binary classification problem, if there are more labels, we have a multiclass classification problem, e.g., assigning blood samples to the blood types "A", "B", "AB" and "O".

A classifier learns a function f that maps an input x to an output y, as shown in equation 4.1. Sometimes, the function f is referred to as classifier instead of the algorithm that implements the classifier.

$$y = f(x) + \varepsilon \tag{4.1}$$

where

f = Function that maps x to y, learned from labeled training data

x = Input, independent variable

y = Output, dependent variable, predicted from unlabeled testing data

ε = Epsilon, irreducible error

ε is the irreducible error that stems from noise and randomness in the training data and that, as the name suggests, cannot be reduced during training. It can be reduced in some cases through more data preprocessing steps, however, ε is a theoretical limit of the performance of the learning algorithm.

Typical classifiers include Bayesian models, decision trees, support vector machines and artificial neural networks.

4.1.1 Artificial neural networks

Artificial neural networks are inspired by the human nervous system. They encompass a large number of different models and learning methods. Here, we cover some of the widely-used models inorder to show their principal functioning.

A typical neuron, as found in the human body, looks as depicted in Figure 4.1. It consists of dendrites that receive electrochemical stimulation from upstream neurons through synapses located in different places on the dendrites. Presynaptic cells release neurotransmitters into the synaptic cleft in response to spikes of electrical activity known as action potentials. The neurotransmitter stimulates the receiving neuron which, in turn, creates an action potential. The action potential is transmitted along the cell membrane down the axon to the axon terminals where it triggers the release of neurotransmitters. The neuron is said to "fire".

The dendrites can receive signals from more than one upstream neuron. Each neuron is typically connected to thousands of other neurons. It is estimated that there are about 100 trillion (10^{14}) synapses within the human brain [25]. Also, synaptic connections are not static. They can strengthen and weaken over time as a result of increasing or decreasing activity, a process called synaptic plasticity. Neurologists

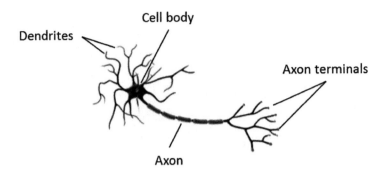

Figure 4.1: Neuron.

have discovered that the human brain learns by changing the strength of the synaptic connection between neurons upon repeated stimulation by the same impulse [37]. When two neurons frequently interact, they form a bond that allows them to transmit the signal more easily and accurately (Hebb's rule). Strong input to a postsynaptic cell causes it to traffic more receptors for neurotransmitters to its surface, amplifying the signal it receives from the presynaptic cell. This phenomenon, known as long-term potentiation (LTP), occurs following persistent, high-frequency stimulation of the synapse. For instance, when we learn a foreign language, we repeat new words until we do not have to concentrate on the translation anymore. We subconsciously use the correct foreign language words. The repeating of the words results in the strengthening of the synaptic connections and formation of a bond between the neurons involved in language speaking. The strengthening and weakening of synaptic connections is imitated in artificial neural networks by linearly combining the input signals with weights. The weights are usually represented in a weight matrix W. Learning in an artificial neural network consists of modifying the weight matrix until the generative model represents the training data well [25].

Interconnected neurons form a neural network. In a similar fashion, artificial neural networks are interconnected neurons that form a network. There are many different types of artificial neural network. One of the simplest forms of artificial neural networks is shown in Figure 4.2. It consists of an input layer, a hidden layer and an output

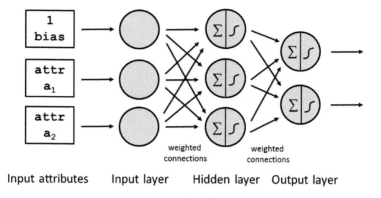

Figure 4.2: Artificial neural network.

layer. Since it has two output neurons, it can be used for binary classification. In fact, one output neuron sufices for binary classification since it can fire or not fire, i.e., emit 1 or 0. The artificial neural network that is shown in Figure 4.2 can also be used for multiclass classification.

In the artificial neural network shown in Figure 4.2, each input neuron is connected with every neuron in the hidden layer and every neuron is connected to every neuron in the output layer; the layers are said to be fully connected. Since the signal passes from the input layer straight to the output layer, it is called a feedforward neural network. Neurons can also have recurrent connections with themselves, as shown in Figure 4.3, or can have connections with upstream neurons to form a recurrent neural network.

An artificial neural network learns a function f that maps an input to an output by adjusting the weights in the weight matrix. Artificial neural networks are also called universal function approximators since they can learn any continuous function with just one hidden layer. In recent years, deep learners (DL) gained a lot of momentum since they perform exceptionally well in tasks such as natural language processing or object recognition. Deep learners are not very different from shallow artificial neural networks in their inner workings. There is no clear definition in literature as to when an artificial neural network is a shallow learner and when a deep learner. Deep learners are, in essence, arti-

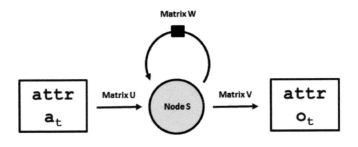

Figure 4.3: Recurrent neural network.

ficial neural networks with many hidden layers. Like artificial neural networks, deep learners learn a function f, but they perform extremely well when the data is very complex. Real world data is often noisy, which can result in f being complicated. Natural language problems, such as machine translation, use large vocabulary sizes. Image analysis requires the processing of large pixel arrays, which can result in many features. In these situations, deep learners often perform much better in learning f from shallow learners.

It should be noted that artificial neural networks and deep learners are simplifications of actual neural networks in living human beings. Real neural networks are highly interconnected, they send signals forward and backwards between layers. Neurons are also not unidimensional processing units but have branches in all directions and they process signals non-digitally as well as digitally. However, the most noticeable difference is that real neurons do not output a single output response, but a spike train, that is, a sequence of pulses, and it is this spike train that encodes information [25].

Advantages of artificial neural networks include their high tolerance of noisy data as well as their ability to classify patterns on which they have not been trained [10]. They can be used when there is little knowledge of the attributes and the relationship between them. Artificial neural networks have been applied to a wide range of real-world problems in Biology, Medicine, Seismology, character and voice recognition or object recognition, to name a few.

Perceptron

We have already discussed the basic functioning of a single neuron, the perceptron, in section 2.1.1. A perceptron, as shown in Figure 4.4, consists of input signals $x_0, x_1, ... x_1$ that are combined with weights w. The inputs can be actual observations or, in the case of multiple layers, intermediate values from upstream perceptrons. The weighted input signals are summed up and passed to an activation function. If the activation function is sufficiently stimulated, it fires, i.e., it emits a signal.

There are different types of activation functions. Figure 4.5 shows some popular types of activation functions. Ramp functions have become the preferred activation functions for deep learners since they simplify the math considerably. An activation function represents the action potential of a real neuron. In its simplest form, it is a step function that either fires, if sufficiently stimulated, or not. It is binary since it outputs either 1 or 0 and can only linearly separate classes. The perceptron calculates $\hat{y}$ by adding up the weighted sum of its inputs and adding the bias.

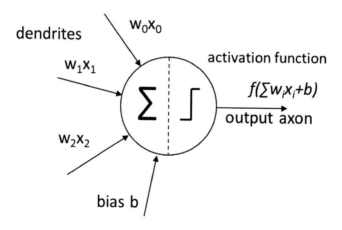

Figure 4.4: Perceptron.

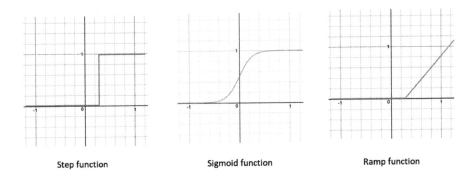

Step function   Sigmoid function   Ramp function

Figure 4.5: Activation functions.

$$f(x) = \hat{y} = \begin{cases} 1, & \text{if } w_0 x_0 + w_1 x_1 + w_2 x_2 + b > 0 \\ 0, & \text{if } w_0 x_0 + w_1 x_1 + w_2 x_2 + b \le 0 \end{cases} \quad (4.2)$$

The input node transmits the signal to the output node without any transformation. The output node sums up the weighted inputs and adds the bias. The bias b is a constant and works similar to the intercept. It shifts the activation function along the x-axis to the left or the right.

As shown in equation 2.3, the input signals are linearly combined with the weights w and the bias b. Because of this, a perceptron with a step function can only linearly separate classes. The line that separates the instances into classes -1 and 1 is called a decision boundary. It is a linear hyperplane. If we need non-linear separation, more than one perceptrons is needed.

The weights are learned during training using a weight update formula given in equation 4.3.

$$w_j^{k+1} = w_j^k + \lambda (y_i - \hat{y}_i^k) x_{ij} \quad (4.3)$$

where

w^k = Weight for i^{th} input after k^{th} iteration
λ = Learning rate

$\hat{y}_i^k$ = Predicted output

x_{ij} = Value of jth attribute of training instance x_i

The initial weights are usually set with random values. The prediction error is calculated as $(y - \hat{y})$. The new weight is a combination of the old weight and the prediction error as shown in equation 4.3. The learning algorithm goes through as many iterations as needed until the output is converging, i.e., the output of the perceptron correctly classifies the training instances. The learning rate λ is a value between 0 and 1 and controls the amount of weight adjustment per iteration. If the learning steps are too small, i.e., λ is close to 0, the algorithm might take a long time to converge, if it is too large it might never fully converge. Therefore, some implementations also adjust the learning rate during training too.

Multilayer Perceptron

Multilayer perceptrons are a class of feedforward networks that consist of at least three layers: An input layer, one or more hidden middle layers and an output layer, as shown in Figure 4.2. Except for the input layer, the layers consist of perceptrons with nonlinear activation functions. Feedforward networks pass the signal in one direction from the input layer through the middle layer(s) to the output layer without any cycles. This is called forward propagation. The multilayer perceptron can separate data points that are not linearly separable. Typically, a multilayer perceptron uses a sigmoid activation function. Not all neurons within the same neural network need to have the same type of activation function. Typical activation functions in a multilayer perceptron include the sigmoid or logistic activation function, shown in equation 4.4, that is also used in logistic regression, and the hyperbolic tangent, shown in equation 4.5.

$$y(x_i) = (1 + e^{-x_i})^{-1} \qquad (4.4)$$

where

e = Euler's number

where y_i is the output of the ith node and e is Euler's number, after the Swiss mathematician Leonhard Euler. x_i is the input to the node and is the weighted sum of the input signals plus the bias. The sigmoid function is called after its S-shaped form and ranges from 0 to 1, but never reaches 0 or 1 as $x \to \pm\infty$.

$$y(x_i) = \tanh(x_i) \tag{4.5}$$

The hyperbolic tangent function looks similar to the logistic function with its sigmoid shape, but ranges from -1 to 1. It is shown in Figure 4.5. The activation functions linearly combine the input signals with weights then sum them up and add the bias.

The output of a perceptron with activation function f is given in equation 4.6.

$$y(x_i) = f(w_1 x_1 + w_2 x_2 + ... + w_j x_j + b) \tag{4.6}$$

where

$y(x_i)$ = Output of ith perceptron

w_j = Weight of the jth input of the ith neuron

b = Bias

Learning in the neural network means learning the weights. During training, the weights are adjusted until the network outputs the desired results. The weights are adjusted using a mechanism called backpropagation that will be explained in section 4.1.1.

A perceptron can have one or more inputs depending on the number of perceptrons in the preceding layer. Multilayer perceptrons are fully connected, meaning that every perceptron is connected with all perceptrons of the next layer.

Multilayer perceptrons can be used to approximate very complex problems and are also called universal function approximators. Multilayer perceptrons have become less popular since there are much simpler models, such as support vector machines, and deep learners have recently been successfully applied for many problems, including object

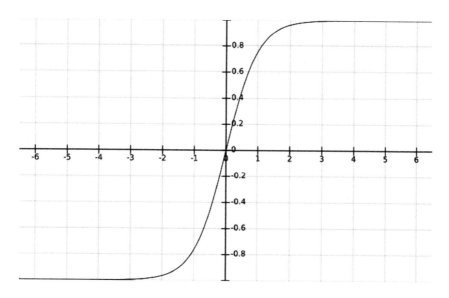

Figure 4.6: Hyperbolic tangent.

recognition, speech recognition and machine translation. Deep learners are not very different from multilayer perceptrons. They also use backpropagation for training but they typically have many more layers and neurons that are not fully connected.

Backpropagation

Backpropagation is a learning algorithm for adjusting the weights during training. After every training iteration, the error or loss at the output of the multilayer perceptron is calculated and propagated back through the network. The weights are adjusted such that the error in the next iteration is minimized. The initial weights are usually set randomly. We do not know what the correct weights of the hidden layer are. This fact gives the neurons in the middle of the network their name; they are called the hidden layer (or layers), because it is not possible to examine and correct their values directly [25]. First, we need to calculate the gradient of the loss or error function at the output. The gradient is calculated using the derivative of the loss function and then the chain rule is applied in order to calculate the gradients of each layer. In every iteration, the gradient is minimized until the minimum of the function is found. This process is called gradient descent. Backpropagation is used by the gradient descent algorithm to adjust the weights. This is sometimes called backpropagation of errors. Gradient descent is an iterative optimization method for finding the minimum of a function. Gradient descent and the issue of getting stuck in a local minimum has been covered in Chapter 2. To calculate the gradient, the loss function needs to be differentiable. The most common loss function in machine learning is the mean squared error (MSE) function 2.9 that was also covered in Chapter 2. By using the square, the bigger the errors are the more they are penalized and the error is always positive. The mean squared error is calculated by taking the difference between the measured, predicted output and the ground truth, squaring it and averaging it across the whole training set. The mean squared error function can be easily differentiated and is often used for regression analysis. Other popular loss functions include the log loss function that is used for classification. As usual, other loss functions and variants exist.

A full training iteration goes through following steps:

1. Forward propagation
   (a) Propagate input forward through the network to calculate the output
   (b) Calculate the error at the output
2. Backpropagation
   (a) Backpropagate the error through the network
   (b) Adjust the weights

A few remarks: A multilayer perceptron can approximate any continuous function and is, thus, a universal function approximator. It can be shown mathematically that only one hidden layer suffices to approximate any function. This is called the universal approximation theorem. This does not mean that a multilayer perceptron always provides meaningful results. The reality is that, sometimes, the more complex neural networks either overfit the data or fail to converge to an optimum solution [1]. In real world applications, we rarely see more than two layers, and, as mentioned before, one layer is enough. However, we do not know how many perceptrons we need. The number of nodes has to be determined through experimentation by training and evaluating networks with different number of nodes.

Multilayer perceptrons have been criticized for being difficult to interpret, particularly their inner workings, unlike decision trees, that are easily interpreted. We know the weights of a trained network and the activation functions, but this does not tell us much. We face the same problem with deep learners too.

The multilayer perceptron is trained using batch training. In every training iteration, the network is presented with the whole training set. The average sum-of-squared errors is calculated and propagated back in order to adjust the weights. The whole cycle is called an epoch, where all the training instances have been passed through the network. A much simpler way to implement training is sequential training, where the error is computed and the weights are adjusted for every training input. This is usually less efficient, but since it is easier to program, we often also see sequential training implementations.

4.1.2   Bayesian models

Bayesian models are based on Bayes theorem. Generally speaking, the Bayes classifier minimizes the probability of misclassification. It is a model that draws its inferences from the posterior distribution. Bayesian models utilize a prior distribution and a likelihood, which are related by Bayes' theorem. Bayes rule decomposes the computation of a posterior probability into the computation of a likelihood and a prior probability [30]. It calculates the posterior probability $P(c|x)$ from $P(c)$, $P(x)$ and $P(x|c)$, as shown in equation 4.7.

$$P(c|x) = \frac{P(x|c)P(c)}{P(x)} \qquad (4.7)$$

where

$P(c|x)$ = Posterior probability of class c (target) given predictor x

$P(x|c)$ = Likelihood which is the probability of predictor given class c

$$
\begin{aligned}
P(c) &= \text{Prior probability of class } c \\
P(x) &= \text{Prior probability of predictor } x
\end{aligned}
$$

The probability in Bayesian models is expressed as a degree of belief in an event that can change in the evidence of new information. The Bayes rule tells us how to do inference about hypotheses from data where uncertainty in inferences is expressed using a probability. Learning and prediction can be seen as forms of inference. Calculating $P(x|c)$ is not easy when $x = v_1, v_2, ... v_n$ is large and requires a lot of computing power. However, Bayesian methods have become popular in recent years due to the advent of more powerful computers.

Naïve Bayes

One of the simplest Bayesian models and one of the simplest machine learning methods is the naïve Bayes classifier. It is a generative probabilistic model and is often a good starting point for an machine learning project since it can be extremely fast in comparison to other classifiers, even on large data sets, and it is simple to implement. In practice, the naïve Bayes classifier is widely used because of its ease of use and despite its simplicity it has proven to be effective for many machine learning problems. For instance, spam filters often use Bayesian spam filtering to calculate the probability that an email is spam or legitimate. It uses word frequencies, a bag of words, as input to detect spam mails and has proved to be an effective filtering method. The naïve Bayes classifier introduces the assumption that all variables v_j are conditionally independent. This is a naïve assumption, hence the name, since variables often depend on each other. This assumption simplifies equation 4.7 to:

$$
P(c|x) = P(c) \frac{\Pi_{j=1}^{n} P(v_j|c)}{P(x)} \tag{4.8}
$$

where

$$
\begin{aligned}
x &= \text{Vector of } n \text{ attributes} \\
P(x|c) &= \text{Likelihood which is the probability of predictor given} \\
&\quad \text{class } c
\end{aligned}
$$

$P(c)$ = Prior probability of class c
$P(x)$ = Prior probability of predictor x

When the assumption holds true, the naïve Bayesian classifier is the most accurate in comparison with all other classifiers [10]. In practice, however, dependencies can exist between variables.

Despite the fact that the independence assumption is often wrong, the naïve Bayesian classifier still performs surprisingly well in real-world applications, even with small training data sets. One advantage is that it returns not only the prediction but also the degree of certainty, which is often very useful. Also, it makes dealing with missing values easy. Due to its simplicity, the naïve Bayesian classifier is less prone to overfitting from artificial neural networks, for example. Besides text analysis, it is also used for weather prediction, medical diagnosis, face recognition and recommendation engines, to name a few.

4.1.3 Decision trees

Decision trees or decision tree induction is an other type of widely-used classifiers. Decision tree induction is the learning of decision trees from class-labeled training tuples [10]. It generates a tree-like structure, where each node represents a decision for an attribute except the leaf nodes. Trees are one of the most common and powerful data structures in computer science in general. Figure 4.7 shows a simple decision tree to predict whether there will likely be a thunderstorm or not, based on the attributes of temperature and humidity.

The top node is called root node and holds the first attribute, the end nodes are called leaf nodes and hold a class label. The branches represent the outcome of a decision on an attribute. In Figure 4.7, we have two binary splits with just "yes" or "no". There can be more than just two splits. In this case, we have a nonbinary tree. For instance, we want to predict what type of car a potential customer is going to buy based on some attributes such as "income", "age", "family status", etc. We might have a split with three branches such as "family car", "sports car" and "luxury car".

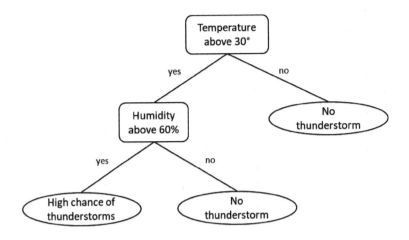

Figure 4.7: Decision tree.

New data navigates through a decision tree starting at the root node and then passing down to the leaf nodes. Each feature is evaluated in a node while progressing to the leaf nodes where we receive the classification result, the class label, of an instance. If the target variable is categorical, such as "family car", "sports car", etc., it is a classification tree, if the target variable is continuous, it is a regression tree. In it's simplest form, a decision tree can consist of just a set of if-then rules (logical disjunctions). Decision trees are flowchart-like and, since decision trees consist of nodes and edges between the nodes, they are essentially restricted graphs.

Developing an algorithm that creates a decision tree is far more complicated than using it. However, most algorithms are based on the same principle. At each step we should choose the attribute that gives us most information based on what we already know. If we want to classify objects, we should ask more general questions such as "is it a vehicle?" before we ask "is it a truck?". Deciding how much information is provided by each question based on certain prior knowledge is a task for information theory and can be challenging. Fortunately, algorithms for decision tree generation, such as ID3, which stands for Iterative Dichotomiser 3, C4.5 and CART, which stands for Classification And Regression Trees, have been around for decades and implementations of modern variations can be found in popular machine learning frameworks, such as WEKA.

Evaluating decision trees differs from other classifiers. We cannot use a loss function since the order of the splits matters in a decision tree. So, we cannot just evaluate the result at the output against a gold standard. When constructing the decision, tree we need to make the better splits early in the tree. We do this by evaluating the information gain at each step. We can use the Gini index or the information entropy to determine the impurity of the data set D, which is a training set of tuples. The Gini index is defined by equation 4.9:

$$D = 1 - \sum_{i=1}^{c} p_i^2 \qquad (4.9)$$

where

p_i = Probability that a tuple in D belongs to class C_i

c = Number of classes

The Gini index, also called Gini coefficient, Gini impurity or Gini ratio, is a measure of statistical dispersion used in economics and has been developed to represent the wealth distribution of the residents of a nation. When building a decision tree, the Gini index is evaluating the splits and lies between 0 and 1. The Gini index considers a binary split for each attribute [10]. If the separation is perfect, the Gini index is 0. If the index is 1, the attributes are equally distributed in each of the two splits. So, we want splits with a low Gini index. The CART algorithm uses the Gini impurity to generate a tree.

Information gain is based on the notion of entropy from information theory. The information gain is the entropy of the parent node minus the entropy of the child nodes. It weights the probability of a class by the log_2 of the class probability. The entropy is defined by equation 4.10:

$$E = \sum_{i=1}^{c} -p_i \log_2 p_i \qquad (4.10)$$

where

p_i = Percentage of each class present in the child node

c = Number of classes

$p_1, p_2, \ldots$ represent the percentage of each class in the child node and add up to 1. The information gain is then the entropy of the parent node minus the entropy of the child nodes. The smaller the entropy value the better, in other words, the difference between the parent nodes entropy is larger. The Gini index caps at 1, whereas the maximum value for entropy depends on the number of classes. It is based on log_2, so for two classes, the maximum entropy is 1, for four classes it is 2, and so on.

Decision trees are very popular classifiers for several reasons. Decision trees are intuitive and easy to interpret. They are often used with text features instead of numeric features, making them human readable. The computational cost of making the tree is fairly low, and the cost of using it is even lower: $O(\log N)$, where N is the number of datapoints [25]. The construction of decision tree classifiers does not require any domain knowledge or parameter setting, and, therefore, is appropriate for exploratory knowledge discovery [10]. A decision tree can be represented by a set of logical disjunctions that can be easily programmed.

Random forests

Random forests are a class of ensemble learners that combine several decision tree models to form a stronger learner from just using a single weak decision tree. It is a divide-and-conquer approach, used to improve the performance of individual decision tree models. Random forest ensemblers average the result of the individual decision trees. In simple words, random forests create several decision trees and merge them in order to get a better prediction. Random forests are usually trained using bagging. In bagging, several models, here decision trees, are trained on subsets of the training data where the data points in the subsets are selected randomly with replacement (bootstrapping). Bagging improves the prediction performance by reducing variance. Decision trees are sensitive to the training data. When the training data changes, the resulting decision tree can be very different and, hence, the result can be quite different too. Bagging improves the result by averaging the prediction from each model. For instance, if we have five decision trees, each trained with a different training data subset, where each tree produces a weather forecast: "sunny", "sunny", "cloudy", "rainy", "sunny", we take the most frequent prediction: "sunny". The bagging algorithm works as follows:

1. Create random subsets of the training data set with replacement, meaning the same data can be reused
2. Train each decision tree with the random subsets
3. Using a new data set, calculate the average prediction of all decision trees

Using bagging, we are not concerned with the individual trees' bias and variance. Typically, they have high variance and low bias. The number of trees needs to be determined through experimentation, for instance, by increasing the number of trees in each run until there is no improvement in the predictive performance anymore. Bagging can be used for classification and regression just like decision trees. Bagging will be described in more detail in Chapter 9.

A problem with bagging is that the trees are often very similar and their predictions are, therefore, highly correlated. Ensemblers work better if the results are not or only weakly correlated. Random trees change how the sub-trees are trained in a way that the results of the individual trees are not correlated. The idea behind random forests is simple. The CART algorithm looks through all variables when deciding on the best feature to split. The random forests algorithm randomly selects a subset of variables from which it can choose the best feature for the split. The number of random features n needs to be defined beforehand and determined by experimentation. A good starting point for classification is $n = \sqrt{p}$ where n is the number of random features and p the total number of features, and $n = \frac{p}{3}$ for regression. Generally speaking, the smaller n the smaller the inter-tree correlation and the strength of the

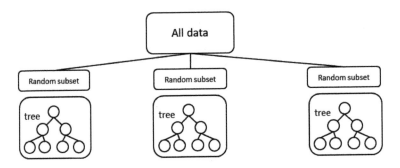

Figure 4.8: Random forest.

individual tree goes down. Figure 4.8 shows a random forest consisting of three decision trees.

The instances that have not been used for training, the left out samples, can be used to estimate the performance of the random forest. Random forests can be used for classification and regression. They can handle missing or unbalanced data. They are less prone to overfitting than other machine learning algorithms. Random forests are almost always better than decision trees. Typical applications of random trees are fraud detection systems in banking transactions, recommendation engines and feature selection.

4.1.4 *Support vector machines*

A support vector machine (SVM) is a widely-used machine learning algorithm that can be applied for classification and regression problems. It is a discriminative, non-probabilistic binary linear classifier. It is often preferred over other machine learning algorithms, such as neural networks, since support vector machines are simpler, less computation is required and yet a high accuracy can often be achieved. There are also variants for non-linear classification. The basic idea behind support vector machines is to find a hyperplane in an n-dimensional space that distinctly separates the data points in two classes, where n is the number of features. In machine learning, a hyperplane is a decision boundary that is used to classify data points. If the data is linearly

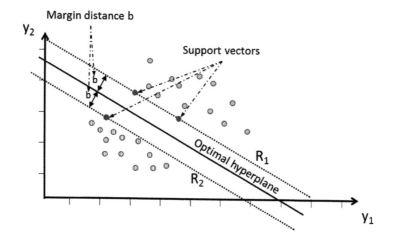

Figure 4.9: Support vector machine.

separable, we can find two parallel hyperplanes that separate the two classes of data. In Figure 4.9, the two hyperplanes, R_1 and R_2, are represented by the two dotted lines. The hyperplane that lies in the middle of the two dotted hyperplanes is the maximum margin hyperplane. The support vector machine algorithm finds the hyperplane with the maximum distance from the nearest training samples. The data points on the two dotted hyperplanes are the support vectors. They influence the position and orientation of the hyperplane. In Figure 4.9, the support vectors are shown as solid dots. They are at the nearest distance b from the hyperplane. The hyperplane is learned using a procedure that maximizes the margin, i.e., that finds a plane with the maximum distance between the data points of both classes. We use the support vectors to maximize the margin of b. The support vector machine algorithm is, thus, called a maximum margin classifier . The larger the margin the better the generalization.

In a two dimensional space, a hyperplane is just a line. A hyperplane has one dimension less than the ambient space. Using support vector machines, we have a set of input features $x_1, x_2, ... x_n$ and a set of weights w_i that we linearly combine with the features, to predict y. This is identical to neural networks. However, using support vector machines, we optimize the margin by just using the weights for the

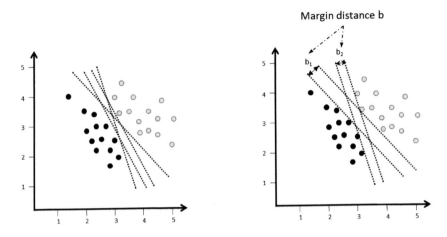

Figure 4.10: Support vector machine hyperplanes.

support vectors. In other words, we only use the features that matter for deciding on the position and orientation of the hyperplane. This is the main difference to neural networks. If we have two dimensions, as in Figure 4.10, we try to find the coefficients a, b, c such that $ax + by \geq c$ for the solid dots and $ax + by < c$ for the pale dots. Looking at the left coordinate system in Figure 4.10, we see that there are many possible solutions for a, b, c and, hence, many possible hyperplanes, but none might be optimal. In the right coordinate system, we can see that, depending on the position of the hyperplanes, the margin can be wider, b_1, or narrower, b_2, depending on the position.

Contrary to linear regression or neural networks, where all the data points are considered, using support vector machines only the support vectors of the training set influence the position of the dividing hyperplane. The support vectors just touch the boundary of the margin. Finding the optimal hyperplane is an optimization problem. The margin is calculated as the perpendicular distance from the support vectors to the hyperplane. The hyperplane is defined by equation 4.11:

$$w^T x + b = 0 \qquad (4.11)$$

where

w^T = The weight vector

x = The input vector

b = Bias

For a given weight vector w and a bias b we can then write:

$$w^T x + b \geq 0 \quad for \quad d_i = +1$$
$$w^T x + b < 0 \quad for \quad d_i = -1$$

where

d = Margin of separation

The margin d is the separation between the hyperplane and the closest data point. The maximal margin is the optimal hyperplane. In a real world setting, the data is typically noisy and cannot be perfectly separated with a hyperplane. The maximum margin rule is, hence, relaxed by introducing additional coefficients called slack variables. This allows for some data points to end up on the wrong side of the hyperplane. This is called the soft margin classifier. However, introducing new coefficients increases complexity and requires more computing power.

As we have seen before, data is often not linearly separable. The data points in the left graphic in Figure 4.11 cannot be separated using a line. To solve this problem, we can use the kernel trick. We use a function φ to map the data into a different space. To do this, instead of using the data points themselves, we use the inner product of any two given data points. Using φ, instead of computing the dot product $x_i \cdot x_j$, we will have to compute $\varphi(x_i) \cdot \varphi(x_j)$, where x_i and x_j are the two feature vectors. This is computationally expensive. Using a kernel function K, such that $K(x_i, x_j) = \varphi(x_i) \cdot \varphi(x_j)$, we do not need to compute or even know φ. In other words, the kernel function defines the inner products in the transformed space.

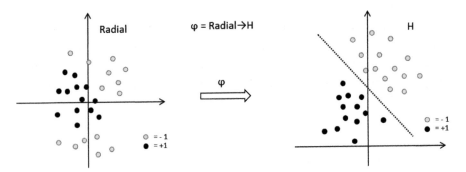

Figure 4.11: Kernel trick.

The kernel defines the similarity of the new data points and the support vectors in the transformed space. The dot product is the similarity measure. Kernels generalize the notion of "inner product similarity".

There are many different types of kernels. Polynomial and radial kernels transform the input space into higher dimensions. Polynomial kernels do not need the dot product. Polynomial kernels can be written as:

$$K(x,x_i) = (x \cdot x_i + 1)^p \qquad (4.12)$$

where

p = Defined a priori

The power p needs to be defined by the user. If $p = 1$ it is a linear kernel. Polynomial kernels allow for curved hyperplanes in the input space.

A radial kernel can be written as:

$$K(x,x_i) = exp(-\frac{||x - x_i||^2}{2\sigma^2}) \qquad (4.13)$$

where

σ^2 = The width defined a priori

This is a radial basis function. The output of a radial basis function depends solely on the distance of the input from the origin. The width σ is usually between 0 and 1 and can be thought of as the spread of the decision region. A radial kernel can create complex regions within the feature space. Radial basis functions are also occasionally used as activation function in neural networks.

Learning a support vector machine is an optimization problem. We can use stochastic gradient descent to search for the coefficients of the hyperplane. However, there are more efficient techniques, such as the sequential minimal optimization method. Support vector machines have been extensively studied and many papers and books on the subject exist. Here, we can only cover the basic concepts.

Support vector machine have many applications. Due to their versatility and the fact that they are usually simpler, they have fewer hyperparameters and they are often preferred over neural networks. They are widely used in biology and other sciences for classification tasks. They also have been applied to text classification problems, image classification and hand-written character recognition. Kernels cannot only be defined over vectors but also over trees, strings, etc. Although the training time of even the fastest support vector machines can be extremely slow, they are highly accurate, owing to their ability to model complex nonlinear decision boundaries [10]. One obvious advantage of artificial neural networks over support vector machines is that artificial neural networks can have any number of outputs, whereas support vector machines have only one. However, support vector machines as well as neural networks are both black boxes and are not as intuitive to understand as decision trees, for example.

4.1.5 *k-nearest neighbor*

k-nearest neighbor (k-NN) is a widely-used classification technique that can also be used for regression. It is widely used since the output can be easily interpreted and it requires low computing time. k-nearest neighbor falls into the category of lazy learners. Contrary to eager learners, such as support vector machines or decision trees, learning does not happen as soon as training instances appear, but only after new instances appear. Eager learners create a generalization from test instances. Lazy learners, such as k-nearest neighbor, compare each new instance with existing ones using a distance measure, delaying learning until new instances are seen by the learner. In other words, learning happens at prediction time. This is also called instance-based learning and requires the test instances to be deployed with the model. It is a completely different way of representing the "knowledge" from the training data since it uses the instances itself and not a rule set or decision tree inferred from the training instances. A new instance is assigned to the class of the closest existing instance. This is called nearest-neighbor classification. Typically, the Euclidean distance is used. Instance-based learning is also called memory-based learning since the training instances are stored in memory. k-nearest neighbor is one of the simplest machine learning algorithms. If it is used for classification, the output is a class membership, i.e., a discrete value. The new instance is classified by a majority vote of its neighbors. If it is used for regression, the output is the value for the object, i.e., it predicts continuous values. k-nearest neighbor uses feature similarity to determine the distance between the training and the new instance.

Sometimes more than one neighbor is used and the average (median) of the distances is used. It is then called k-nearest neighbor, where k is the number of neighbors. In the left quadrant of Figure 4.12, $k = 3$ and the new star-shaped instance belongs to the class of the light grey instances. On the right quadrant, if $k = 3$, the new instance also belongs to the class of the light gray instances since 2 of the 3 closest neighbors are light gray and only one belongs to the class of dark grey instances. If $k = 5$, the new instance belongs to the class of the dark grey instances, since 3 of the closest instances are dark grey. So choosing the parameter k is crucial in this algorithm.

The complexity of instance-based learners grows with the training data. The more training instances we have the more computing power is required. One advantage of instance-based learners is their ability to adapt to new data. They simply store new training instances without the need to retrain a model.

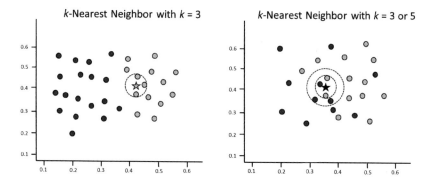

Figure 4.12: k-nearest neighbor.

k-nearest neighbor usually stores a subset of the training instances in order to reduce complexity, and not the whole training data set. The training set is divided into several classes. A new data point is assigned to a class based on feature similarity. The assignment is calculated by a majority vote of its neighbors. The predicted class is the most common among its k neighbors. If k-nearest neighbor is used for regression, the average of the values of its k nearest neighbors is calculated.

As mentioned before, choosing k is crucial for the k-nearest neighbor algorithm since different values for k might yield different predictions. Since k is based on majority voting, a small k makes the prediction less stable since only a few instances decide on the class assignment. As k gets bigger, the decision boundary becomes less distinct. Finding an optimal value for k is not an easy task and requires some trial and error. We can start with a low k and determine the accuracy of the prediction. We then increase k until the accuracy does not improve anymore or gets worse. It is good practice to have an odd number for k, otherwise we might end up with a draw, e.g., for binary classification 2 votes for class A and 2 votes for class B.

A few remarks: k-nearest neighbor is a non-parametric, lazy learning algorithm. Non-parametric means the algorithm does not make any assumptions on the underlying data distribution. This makes k-nearest neighbor a good choice if there is no prior knowledge about the underlying data distribution. Real-world data does not follow theoretical assumptions which makes it a "natural" choice for many non-linear data sets. k-nearest neighbor is also simple to understand and easy to interpret but still provides high accuracy which makes it a good choice of starting point for a machine learning project. On the down side, since k-nearest neighbor stores all or almost all training data, it requires a lot of memory and computing power, especially when k is big.

A typical application of k-nearest neighbor is market segmentation. People with similar features such as age, income, family status, etc., are likely to have similar purchasing behavior. So, k-nearest neighbor can be used for recommendation engines and marketing campaigns. However, if the data volumes get too large, such as the Netflix or Amazon data sets that are enormous, k-nearest neighbor is not feasible anymore since every new instance has to be compared with all the training instances. Determining creditworthiness is another application of k-nearest neighbor. People with similar financial features are given the same credit rating.

4.2   Regression analysis

Regression analysis is a set of statistical methods applied to estimate the relationship among variables. It is used to determine if a correlation exists between variables. Regression analysis is probably one of the most important data analysis methods. Linear and logistic regression are the most popular regression analysis algorithms in machine learning, but many more exist.

Regression analysis is used to make data driven decisions, such as forecasting, e.g., predicting sales volumes, time series analysis, e.g., development of share price over a period of time, or correlation analysis. However, it is important to notice that correlation does not imply causation. If a regression line fits a set of data points well, this does not mean that there is a cause-and-effect relation. For instance, on a rainy day, people tend to go shopping more than on a sunny day. There is a correlation between weather and sales volumes, but that does not mean that bad weather causes the sales.

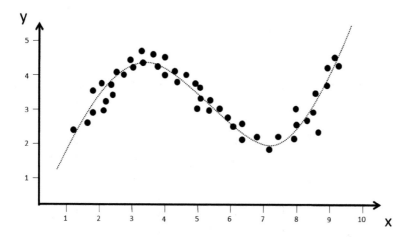

Figure 4.13: Regression analysis.

In essence, regression analysis aims to fit a line or curve to a set of data points in such a way that the distance of the line or curve to the data points is minimized. Figure 4.13 shows a regression curve fitted to a set of data points.

Some remarks about Figure 4.13:

■ Not all points are on the curve, some might even be distant from the regression line

■ The regression line is an approximation

■ It indicates if there is a significant relationship between the dependent and independent variable

■ The effects of the measured variables can be compared using different scales

Regression analysis is a statistical approach for understanding the relationship between input and output variables. In machine learning we are more concerned about making accurate predictions by minimizing the error of a model. We use regression to determine:

1. how well a set of predictor variables predict an outcome, the dependent variable(s)
2. which variables in particular are significant for the outcome

There are many different kinds of regression techniques available to make predictions. Also, new regression methods can be created. In Figure 4.13, a curve has been fitted, since a straight line would not provide a good fit for the data points. Which regression method is most suited depends mostly on three factors:

■ The number of independent variables
■ The type of the dependent variables
■ The shape of the regression line

Regression analysis is used to explain a phenomenon, for instance, why did sales numbers drop last month, in order to make a prediction about the future, for instance, the risk of heart disease in relation to daily number of cigarettes smoked, or to support the decision making process, for instance, which promotion will have the biggest effect on sales volumes. Regression analysis lets us determine mathematically, which variables have an impact, which factors matter most, which ones can be ignored, how the factors are correlated, and, most importantly, how certain we are about these factors.

Next to linear and logistic regression, there are other types, such as polynomial regression, lasso (least absolute shrinkage and selection operator) regression and stepwise or ridge regression, to name a few. Regression models can also be combined if needed. Typically, a linear model is used first. If no adequate fit can be found, a nonlinear model can be tried, since nonlinear models can fit a wider variety of curves.

As with all machine learning schemes, the first step in regression analysis is collecting data. For instance, if we want to predict sales volumes in the next quarter, we first collect sales volumes over the past year and plot the data as shown in the left quadrant of Figure 4.14.

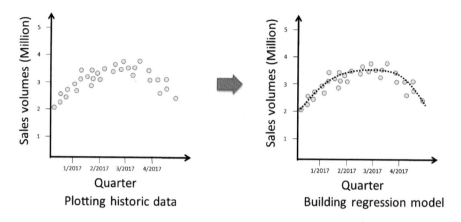

Figure 4.14: Regression analysis steps.

The *y* axis shows the sales volumes, the dependent variable, and the quarter on the *x* axis. The *y* axis is always the dependent variable, the variable we want to predict. Each data point represents one week's data. The curved line in the right quadrant summarizes the relationship between sales volumes and the quarter of the year. The sales goods could be grillables that are typically sold more in summer than in winter.

As with all machine learning algorithms, overfitting is a problem in regression analysis. When the input data is highly correlated, linear regression will overfit. Other regression models, such as polynomial regression, also tend to overfit, so data regularization techniques must be used to prevent nonsensical results. Figure 4.15 shows an overfitted regression curve. An overfit learner captures noise and does not generalize well on new, unseen data. Some regression models, such as lasso or ridge regression, have been developed, to mitigate the problem.

4.2.1 Linear regression

Linear regression is a statistical method of studying the relationships between two continuous variables. It is one of the simplest methods

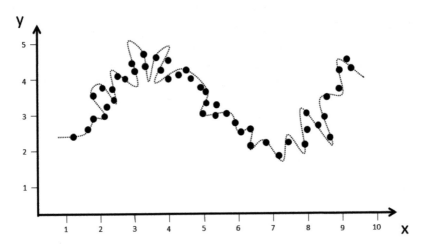

Figure 4.15: An overfitted regression curve.

used for machine learning. Linear regression attempts to estimate the relationship between scalar dependent and independent (explanatory) variables by fitting a linear equation to observed data. If one explanatory variable is present, it is called simple linear regression, if more than one explanatory variable is present, it is called multiple linear regression.

Linear regression is a basic approach used for predictive analysis. It is a linear approach for modelling the relationship between the dependent variable y and one or more explanatory or independent variables x. The linear equation 4.14

$$y = ax + b \qquad (4.14)$$

where

x	=	Independent Variable
y	=	Dependent Variable
a	=	Slope
b	=	Intercept

is the simplest form of the regression equations with one dependent and one independent variable. a and b are the coefficients, also called regression coefficients or weights. Linear regression assumes a linear relationship between the input variable or variables x and the single output variable y. Using machine learning, the coefficients are learned from data. In other words, we try to find the best fit line for the given data points. A variable is called feature in machine learning. The dependent variable is called target, predictor or label in machine learning terms.

Figure 4.16 shows the regression line that has been fitted to the data.

If the data cannot be fitted linearly, a polynomial or curved line might better fit and techniques such as polynomial or curvilinear regression might apply.

The equation that describes how y relates to x is called regression model. It is given in equation 4.14. It calculates the mean value of y for a known value of x. During training, the coefficients a and b are adjusted until a best fitting line is found, the regression line. This is achieved by minimizing the sum of the squared difference of the distance between the data points and the regression line.

First, we need to determine, if there is a relationship between the dependent and the independent variables. For instance, we might want to

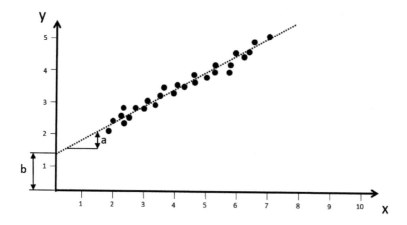

Figure 4.16: Linear regression.

determine if there is a relation between body weight and life expectation. If a relationship exists, we can predict the life expectation for new weights.

The most common method for fitting a linear regression line is the ordinary least squares method. This is also called ordinary least squares linear regression or just least squares regression. Ordinary least squares estimates the values of the coefficients by calculating the distance of each data point from the regression line, squaring it and adding all the squares together. Ordinary least squares tries to minimize the summed up squared errors. Variations on the ordinary least squares regression are lasso (least absolute shrinkage and selection operator) regression and ridge regression. Lasso regression also aims to minimize the sum of the squared error, but it also performs feature selection and minimizes the absolute sum of the coefficients (Lasso regularization). Ridge regression also minimizes the squared error, additionally in minimizes the squared absolute sum of the coefficients. Lasso and ridge regression are two popular regularization methods that can prevent overfitting.

Simple linear regression

In simple linear regression, also called bivariate regression, an observation consists of pairs of variables, one independent, x, and one dependent variable y. A straight line expresses the relationship between the two variables. The best fit for the regression line is expressed using the least square approach. The best fit line is the one where the prediction error or residual for each data point is as small as possible. The residual is the vertical distance between the observer data point and the regression line. If the prediction error is not squared, the positive and negative errors would cancel each other out.

The relationship between x and y is expressed in equation 4.15:

$$y = ax + b + \varepsilon \qquad (4.15)$$

where

x	=	Predictor Variable
y	=	Dependent Variable
a	=	Slope
b	=	Intercept
ε	=	Error term

The error term ε or regression residual represents the fact that regression is an approximation and not perfectly precise. The error term is needed in order to correct for a prediction error between the observed and predicted value. It is also called residual, disturbance or remainder term and is created when the model does not fully represent the actual relationship between the independent variables and the dependent variables. It is the sum of the deviations with the regression line and it is an observable estimate.

If the regression line slopes up we have a positive relationship, if it slopes down we have a negative relationship. A vertical regression line means there is no relationship between the variables. Some examples of possible relationships are hours of study and test mark, alcohol consumption and blood alcohol level or years of experience and income.

Multiple linear regression

Simple linear regression uses only one predictor variable to predict the factor of interest. In many real-world problems, more than one independent variable influences the prediction, so, usually more than one predictor is considered for regression analysis. If two or more independent variables exist, it is called multiple linear regression or multivariable linear regression. Multiple linear regression attempts to model the linear relationship between the the explanatory (independent) variables and the response (dependent) variable by fitting a linear equation to observed data points.

Given n observations, the model can be defined as shown in equation 4.16:

$$y = \beta_0 + \beta_1 x_1 + \beta_2 x_2 + ... + \beta_n x_n + \varepsilon \qquad (4.16)$$

where

x	=	Predictor Variable
y	=	Dependent Variable
β	=	Regression coefficient
n	=	Number of observations
ε	=	Error term

Multiple linear regression assumes that:

- There is a linear relationship between the dependent variable and the independent variables
- The independent variables are not too highly correlated; absence of multicollinearity
- The regression residuals are normally distributed

Since we have more than one independent variable x, we are fitting a hyperplane into a multidimensional space. If we have n independent predictor variables we try to fit a hyperplane into a $n + 1$-dimensional space. To find the best fitting hyperplane we can use the same method as for simple, linear regression and minimize the sum of squared residuals. However, with high dimensionality the calculations can become quickly complicated and optimized methods using linear algebra are employed to make the computation more efficient. Also, since regression analysis provides an estimation, we need to test the model for significance. ANOVA (Analysis of variance) is a set of statistical methods that can be used to test the significance of the parameters. However, these methods are described extensively in literature and are beyond the scope of this book.

Some examples of relationships where more than one independent variable is used are estimating house price that depends on the location, the size of the house, its age etc., or life expectancy that depends on body weight, sporting activities, medical history etc.

4.2.2 Polynomial regression

Polynomial regression can be used if there is a nonlinear relationship between the variables x and y. The relationship between the dependent and the independent variable y and x is modeled using a nth degree polynomial. For one single predictor x, the model can be defined as shown in equation 4.17:

$$y_i = \beta_0 + \beta_1 x + \beta_2 x^2 + ... + \beta_n x^n + \varepsilon \qquad (4.17)$$

where

x	=	Predictor Variable
y	=	Dependent Variable
β	=	Regression coefficient
n	=	Degree of the polynomial
ε	=	Error term

Polynomial regression can fit a nonlinear model, but since the regression function is linear, it is considered a special case of multiple linear regression. Even though x^2 is quadratic, the unknown coefficients β_0, β_1 ... are linear and the estimator considers x^n just as another variable. We can use the same analysis procedures such as least squares just as in multiple linear regression.

The equation 4.17 is relatively simple but polynomial regression models can have many more predictor variables. Finding the order of the polynomial is not straight forward. As a rule of thumb, the degree of the polynomial should not exceed half the sample size. It is good practice to plot the data and the fitted curve, then use human judgement to assess the subjective goodness of fit: determine "practical significance" versus "statistical significance".

A few guidelines to consider when fitting a polynomial regression model are:

■ You should not extrapolate beyond the limits of the sample data points

■ If the model includes x^n and x^n shows to be a statistical significant predictor, you should also include x^m where $m < n$ even if x^m is less significant

■ If the resulting plot of the curve looks reasonable use statistical significance only to support a model

4.3 Logistic regression

Logistic regression (LR), or logit regression, is used to estimate the binary dependent variable based on one or on a set of discrete explanatory values. It is a widely-used model in machine learning and is used in many different areas. Logistic regression does not need a linear relationship between the dependent and the independent variable. It is used for classification when the dependent variable is binary (dichotomous), e.g., 0/1 or true/false. This is why it is also called binary logistic regression. Logistic regression calculates the probability of an event, such as success/failure. If the dependent variable has more than two categories, it is called multinomial logistic regression.

At the core of logistic regression is the logistic function. The logistic function or logit function is a sigmoid curve. A sigmoid curve is an S-shaped curve between 0 and 1 that never reaches 0 or 1 as shown in Figure 4.17. Mathematically the logit function is defined as shown in equation 4.18.

$$logit(x) = \frac{1}{1 + e^{-x}} \tag{4.18}$$

where

x = Independent variable

e = Euler's number

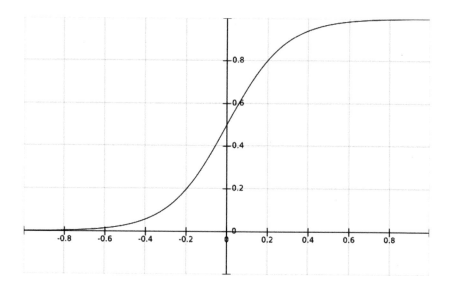

Figure 4.17: Logistic regression curve.

The logit function maps any real number x into a value between zero and one. In logistic regression the explanatory variable x is linearly combined with coefficient values β_0 and β_1. These coefficients are learned during training. The logit function can then take the form as shown in equation 4.19.

$$p(x) = \frac{1}{1 + e^{-(\beta_0 + \beta_1 x)}} \qquad (4.19)$$

where

$\quad x \quad = \text{Explanatory variable}$

$\quad e \quad = \text{Euler's number}$

$\quad \beta_0 \quad = \text{Intercept}$

$\quad \beta_1 \quad = \text{Regression coefficient}$

p is interpreted as the probability of the dependent variable belonging to a class such as "success" or "failure". Since logistic regression yields a binary prediction, the probabilities need to be transformed into 1 or 0. Ideally, the coefficients β_0 and β_1 are learned in a way that the learner predicts values close to 0 or 1. In practice, we can use the probability directly using a threshold, for example 0.5:

$$y = \begin{cases} 1, & \text{if } p \geq 0.5 \\ 0, & \text{otherwise} \end{cases}$$

x can be any positive or negative value to infinity, the resulting probability is always between 0 and 1. x can be a combination of several explanatory variables.

Usually for logistic regression the maximum likelihood estimation is used to learn the coefficients. The maximum likelihood estimation is a common learning algorithm used to optimize the coefficients for a variety of machine learning algorithms. Maximum likelihood estimation uses the likelihood function to maximize the likelihood that the coefficients found for the model describe the data that is actually observed and, thus, minimize the prediction error. Maximum likelihood estimation will be explained in more detail in Chapter 7.

Logistic regression examines questions such as: What is the probability of getting lung cancer for every additional pack of cigarettes smoked per day. Some points to consider with logistic regression are:

- Logistic regression can be difficult to interpret
- Outliers should be removed from the data set
- There should be no high correlations (multi collinearity) among the predictors, otherwise the expected likelihood estimation process that learns the coefficients might fail to converge
- To perform well, logistic regression needs large sample sizes
- Logistic regression assumes no error in the output variable y, misclassified instances should, thus, be removed from the training set

Other variants of the logistic regression equation from the one shown in equation 4.19 are possible. The logit function is also used in artificial neural networks as activation function.

Logistic regression has become the preferred method for many binary classification problems. Logistic regression is widely used in medicine to predict mortality of patients or the severity of an illness. It is also used in life science or in marketing to predict customer churn or the success of a marketing campaign. Is is also used for earthquake prediction or to find new oil sources based on seismic activity.

Chapter 5

Evaluation of Learner

CONTENTS

5.1   Evaluating a learner

Learning can be defined as improving performance based on past data. To make sure the learner improves during training, we need to evaluate the learners performance during training. We also need to make predictions about how well a trained learner will perform on new, unseen data. A trained learner needs to accurately capture the signal in the training set, and, at the same time, generalize well on new data. We could just use the training error to measure the learners accuracy, however, we cannot use the training error to predict how well the learner will perform on new data. Also, the training error is not a good estimate of test error, as it does not properly account for model complexity [11].

To evaluate how well a supervised machine learning scheme has learned, we usually use labeled data that was not used for training.

Typically, we divide the data into three data sets: One data set for training, one for evaluating and one for testing the learner. The training set is used to train the learning algorithm, the validation set is used to optimize the parameters of the learning algorithm and the test data set in order to calculate the true error rate on the final, trained algorithm. Each data set needs to be independent in order to obtain a reliable estimate of the error rate. Data sets are typically randomly divided into the three sets. Each data set needs to be representative and no feature should be over- or underrepresented in one set. The proportion of the sets depends on the problem and the amount of available data. Typically the largest amount is used for training and we might use proportions such as 60:20:20 or 70:15:15. The drawback is that this reduces the amount of data that we can use for training. If there is a shortage of data, techniques such as k-fold cross-validation can be used. The data is randomly divided into k data subsets and $k - 1$ data sets are used for training and one holdout data set is used as the validation set. This process is repeated until each of the k subsets has been used as the evaluation set once.

A learner also needs to be evaluated so that it can be compared with other learners. Often, more than one model is trained, and we want to use the learning scheme that best works for our problem. Regardless of whether we train only one learning algorithm or more, we still need to evaluate how good the result is.

There are many different metrics to measure the performance of a predictor and not all can be used for all learners. Some are suited for classification where others might be for regression analysis.

Evaluating a learner is an important task since it gives us an indication as to how well a trained model will perform on new data. In fact, improved evaluation methods have contributed to the recent successes in machine learning and in deep learning.

5.1.1 Accuracy

One metric for evaluating classification models is accuracy. Accuracy is simply the fraction of predictions the model classified correctly. Formally, accuracy is defined as:

$$Accuracy = \frac{Number\ of\ correct\ predictions}{Total\ number\ of\ predictions} \tag{5.1}$$

However, accuracy does not tell the full story. Lets assume we want to classify medical images of tissue samples into cancerogenous and benign tissue samples. Lets assume we have 100 samples, of which 91 are benign and 9 are cancerogenous. If a classifier correctly determined 90 tissue samples as benign and 1 as cancerogenous, using equation 5.1 we end up having an accuracy of $\frac{90+1}{100} = 0.91$. 91% is a very good accuracy, however, only one of the 9 cancerogenous samples was correctly classified as malign. Despite the high accuracy, most malign samples go undiagnosed. In other words, we need to take the number of misclassifications into consideration. Accuracy works particularly poorly in situations where we have a class-imbalanced data set, meaning there is a big difference in the number of positive, cancerogenous, and negative, benign, samples. Accuracy treats all samples the same. However, in real world problems, imbalanced data sets are the rule. A metric that is much more suited for class-imbalanced data sets is precision and recall.

5.1.2 Precision and recall

Precision and recall take into account the correctly as well as the incorrectly classified instances. We use the terms true positives (TP), true negatives (TN), false positives (FP), and false negatives (FN). Positive and negative refers to the classifiers prediction. True positives and true negatives are correctly classified instances. For instance, true positives are correctly classified medical images as cancerogenous and true negatives are correctly classified images as benign. False positives and false negatives are incorrectly classified images. Figure 5.1 visualizes

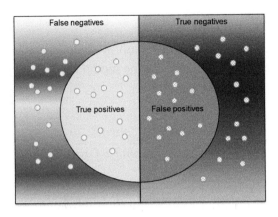

Figure 5.1: Precision recall.

the result of a classification task with the distribution of TP, TN, FP and FN.

For a classification task, precision, also called positive predictive value, is the proportion of correct positive classifications. Precision is defined as:

$$Precision = \frac{TP}{TP+FP} \tag{5.2}$$

Recall, also called true positive rate or sensitivity, is the proportion of actual positives correctly identified. Recall is defined as:

$$Recall = \frac{TP}{TP+FN} \tag{5.3}$$

We can visualize precision and recall by using the fractions of TP, FP and FN, as shown in Figure 5.2.

To fully evaluate the result of a classification task, we need to examine both, precision and recall. In practice, improving precision often reduces recall and vice versa. If the number of false positives decreases, the number of false negatives increase. There is always a tradeoff between precision and recall and it depends on the problem at hand what

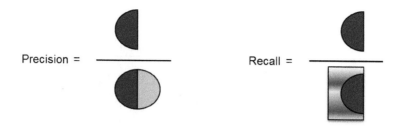

Figure 5.2: Precision and recall.

good values are for both. It is difficult to compare models with high precision and low recall or vice versa.

A metric for the accuracy of a classification task, which considers both precision and recall, is the F-score, or F_1-score. F-score is the harmonic mean of precision and recall and is defined in equation 5.4:

$$F-score = 2 \times \frac{precision \times recall}{precision + recall} \tag{5.4}$$

An F_1-score reaches 1 for perfect precision and recall, and 0 in the worst case. Since the F_1-score uses both, precision and recall, it is useful to compare models with high precision and low recall or vice versa. In practice, F_1-score is often used in natural language processing tasks, such as document classification, or to evaluate the performance of search result classification.

5.1.3 Confusion matrix

A common way of visualizing the result of a classification task is to use a contingency table called a confusion matrix. The rows of the confusion matrix represent the predicted values, the columns represent the actual values. A confusion matrix can have two or more columns depending on the number of labels. A confusion matrix makes it easy to determine if the classifier is confusing two classes, hence its name. For instance, we can visualize the result of the cancer tissue classification using a confusion matrix, as shown in Table 5.1.

Table 5.1: Confusion matrix for the cancer tissue example.

		Actual class	
		Cancerogenous	Benign
Predicted class	Cancerogenous	True Positive (TP): Result: 1	False Positive (FP): Result: 1
	Benign	False Negative (FN): Result: 8	True Negative (TN): Result: 90

If all samples were correctly classified only the fields in the diagonal white cells of Table 5.1 have values, the other fields are all null. A confusion matrix allows a more detailed analysis of a classification result than using mere proportions of classifications and misclassifications since it contains all results.

If there are more that two class labels and we have a multiclass classification problem, the confusion matrix has a row and column for each class label. For instance, if we want to classify wines according to their grape species, we might get a confusion matrix as shown in the upper matrix of Figure 5.3. A confusion matrix also lets us compare one class versus all the other classes, as shown in the lower matrix in Figure 5.3, where Shiraz is compared with all the other grape species.

Multiway	Shiraz	Pinot noir	Merlot
Shiraz	17	3	1
Pinot noir	3	25	2
Merlot	4	2	13

Shiraz vs all	Shiraz	Pinot noir/Merlot
Shiraz	17	4
Pinot noir/Merlot	7	42

Figure 5.3: Multiclass confusion matrix.

There are other variants. A confusion matrix can contain the micro-averaged results which are the sum of the results in the one-versus-all matrices. If we divide the sum of the elements on the leading diagonal by the sum of all of the elements in the matrix, which gives the fraction of correct responses, we get the accuracy [25]. We can then define accuracy as the sum of the number of true positives and true negatives divided by the total number of examples as shown in equation 5.5.

$$Recall = \frac{TP + TF}{TP + FP + TN + FN} \qquad (5.5)$$

We have already seen that accuracy is not giving us the complete picture of how well a classifier performs and we can use precision and recall to help us interpret the result. Another useful metric is the receiver operator characteristic.

5.1.4 Receiver operating characteristic

The receiver operating characteristic (ROC) is a plot of the true positive rate, also called sensitivity or recall, in function of the false positive rate, also called specificity. It is used to measure the performance of a binary classifier at various threshold settings. The true positive rate is the sensitivity or recall as defined by equation 5.3. The false positive rate is $1 - specificity$, where specificity is defined as is equation 5.6:

$$Specificity = \frac{TN}{TN + FP} \qquad (5.6)$$

The false positive rate is then defined as in equation 5.7:

$$FPR = 1 - Specificity = \frac{FP}{TN + FP} \qquad (5.7)$$

For a perfect classifier, the receiver operating characteristic curve will move up straight along the y axis untill it reaches $y = 1$, then along the x axis (100% sensitivity, 100% specificity). If the receiver operating

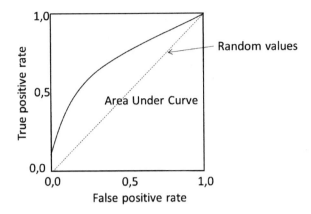

Figure 5.4: Receiver operating characteristic.

characteristic curve is the diagonal, the values are purely random. Each point on the receiver operating characteristic curve represents a sensitivity/specificity pair corresponding to a particular decision threshold. Normally, the curve lies somewhere above the diagonal. The closer the curve is to the upper left corner the higher the accuracy of the classifier. Figure 5.4 shows a receiver operating characteristic curve. It is also called area under the curve (AUC) or AUC-ROC. The area under the curve is the probability that a classifier will score a randomly selected data point higher from a randomly selected negative one. By looking at Figure 5.4, the receiver operating characteristic curve can be used to select a threshold which maximises the true positives, while minimising the false positives.

The receiver operating characteristic curve lets us evaluate the performance of a classifier over its entire operating range. Most often in machine learning, the receiver operating characteristic curve is used to compare statistical models. A single threshold can be used to compare the performance of two or more classifiers at that point, or the entire curve can be compared. A receiver operating characteristic curve of less than 0.5 might indicate that the classifier finds relationships that are the opposite of what we expect and the experiment might have been set up incorrectly.

Using the receiver operating characteristic curve to diagnose classifiers with more than two classes requires more dimensions and complexity increases dramatically. Some efforts have been made using receiver operating characteristic curves for three classes called three-way ROC. Due to the popularity of receiver operating characteristic curves, efforts have been made to use receiver operating characteristic curves to asses other supervised models. Namely, receiver operating characteristic for regression models, called regression error characteristic (REC) curves and regression ROC (RROC) curves, have been proposed.

Chapter 6

Unsupervised Learning

CONTENTS

Unsupervised learning methods are used when there is no labeled data. In other words, there is no ground truth. The most widely used unsupervised method is clustering. Instead of predicting a class as we do with supervised learning, clustering partitions data points into groups based on some similarity measure. In clustering, the goal is to understand data by finding the underlying structure in the data set. Since there is no ground truth in clustering, the evaluation of a cluster is more challenging than evaluating supervised learners. For instance, in a library, books are grouped into fiction, children's books, travel books, cooking books, etc. The similarity is the subject of the books. However, children's books are often fictional too and might belong to more than one cluster. It is not always obvious how to measure the performance of a cluster algorithm and the measurement is often subjective. The clusters have to be useful for the problem at hand. In the library example, clustering is useful if it helps in finding books more easily in a library.

We can use clustering to understand data, but we can also use clustering as a starting point for other data mining tasks. For instance, if a data set is too large to be processed efficiently, clustering can be used for data summarization. Data is grouped in clusters based on similarity to a cluster prototype. A prototype is often a medoid, the most representative object in a cluster. Each cluster prototype is then presented to the data analysis algorithm separately, instead of the whole data set. If many instances in a data set are highly similar, clustering can be used for data compression. Instead of analyzing the whole data set, each cluster prototype is assigned an index that is tabulated. The data is compressed whilst maintaining its structure and usefulness. Compression can be used if a substantial reduction in data size is required and some information loss is acceptable.

Clustering has many applications. It has been applied in biology to group animals and plants into species, genus, families, etc. More recently, it has also been applied in genetics to find similarities in genomes and identify regions with possible pathogenic gene mutations. Other applications include marketing research, where consumers are grouped into potential buyers of certain products or services. Anomaly detection has gained a lot of attention in recent research since it can detect anomalies in network traffic and fend off potential threads and data breaches or help prevent outages in factories based on anomalies in sensor data from machines. Clustering is used for image segmentation to detect regions in an image for object recognition. Clustering is also used for analyzing seismographic data to find new oil sources or detect dangerous zones in earthquake areas. Many more application fields exist. It should be noted that in many real-world applications, the notion of cluster is not well defined which can lead to unwanted clusters or instances ending up in the wrong cluster. For instance, when books are clustered, a biography of a political figure might end up in the politics cluster instead of the biography cluster. Since the number of desired clusters is often unknown at the beginning of a clustering project, there might be one or more unwanted clusters, for instance, a cluster with biographies about political figures. The imprecise definition of clustering led to a wide variety of different cluster algorithms. Also, different clustering methods are categorized differently in literature.

Since clustering groups data into partitions with similar features, it can be considered a classification method. However, in clustering, the labels are derived from the data and often not known beforehand.

6.1 Types of clustering

Clustering can be exclusive, where every instance belongs to exactly one cluster. This is also called hard clustering. Clusters can also be overlapping, where some instances belong to more than one cluster. Figure 6.1 shows an exclusive cluster, where every data point belongs to a single cluster, and a non-exclusive cluster where data points can simultaneously belong to more than one cluster.

In fuzzy clustering, each instance belongs to every cluster to a certain degree. Every data point is assigned a weight or a likelihood that it belongs to a certain cluster.

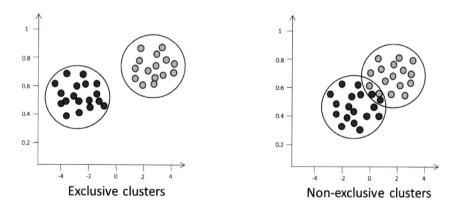

Figure 6.1: Exclusive and non-exclusive clusters.

In a cluster, every data point is closer, or more similar, to any data object in the cluster than to any object in another cluster. Ideally, the clusters are far from each other. This is the notion of well-separateness. A cluster is well-separated if the distance between the data objects in a cluster is small, i.e., the cluster is cohesive, and the distance between data points in different clusters is large. Clusters do not have to be globular, as shown in Figure 6.1, but can have any shape. Also, clusters can have more than two dimensions.

6.1.1 Centroid, medoid and prototype-based clustering

In centroid-based clustering, the data points are grouped around a centroid, the mean of all the objects in a cluster. A centroid usually does not correspond to an actual data point. By contrast, a medoid must be a data point. A medoid is the most representative data point in a cluster. It is used when there is no meaningful centroid, e.g., when the data is categorical and not continuous. The prototype in prototype-based clustering defines the cluster. Each data point is closer or more similar to the prototype than a prototype in any other cluster. The prototype is often the centroid. The prototype can be regarded as the most central point and the clusters tend to be globular. If the clusters have no centroid, the prototype can be the medoid.

6.1.2 Density-based clustering

A cluster is a high density region of instances surrounded by a low density area. The high density area can be of any shape and the clusters can be intertwined. Density-based models are used when the clusters are irregular and when noise and outliers are present.

6.2 k-means clustering

k-means clustering is one of the most widely-used clustering algorithms. We have already introduced k-means clustering in Chapter 2.2.1. It is a prototype-based, partitional clustering algorithm. Partitional clustering means, each instance is assigned to a cluster exactly once, and no instance remains unassigned to clusters [43]. k-means clustering groups instances into k disjoint clusters where k is user defined. A k that has been selected inappropriately can yield poor results.

k-means clustering is an iterative approach. The algorithm starts by randomly selecting k initial centroids from the data set. k has to be selected by the user and is not necessarily known at the beginning of a clustering project. Each instance of the data set is then assigned to a cluster, based on its distance to the centroid. Usually, the Euclidean distance is used. The Euclidean distance works well in a two dimensional setting. However, depending on the data at hand and in higher dimensional environments, other distance measures, such as the Jaccard index or Chebyshev distance, might perform better. After each instance has been assigned to a cluster, the centroids are recomputed by calculating the mean of each instance in a cluster. This process is repeated until the centroids do not change anymore. The basic k-means algorithm is formally defined as described in algorithm 6.1.

Algorithm 6.1 k-means algorithm

1: *Randomly select k initial centroids*
2: *repeat*
3:     *Assign each point to the closest centroid*
4:     *Recalculate the centroids of each cluster*
5: *until centroids do not change anymore*

Often, a weaker convergence criterion is used. The centroids are shifting as long as points move from one cluster to another. Instead of converging when no points move anymore, the steps three and four in algorithm 6.1 are repeated until, for example, more than 1% of the points are moving.

The k-means algorithm can be very time consuming, since a lot of distance measures have to be calculated in every iteration to compute the similarity to a centroid. Many variants to k-means have been proposed in order to improve performance. Especially in low dimensional Euclidean space, many similarity computations can be avoided, thus, significantly speeding up the algorithm. The Euclidean distance is given in equation 6.1.

$$d(X,Y) = \sqrt{\sum_{i=1}^{n}(x_i - y_i)^{-2}} \qquad (6.1)$$

where

$$X = (x_1, x_2, ..., x_n)$$
$$Y = (y_1, y_2, ..., y_n)$$

X and Y are n-dimensional feature vectors in $\mathbb{R}^n$. The goal of clustering is usually expressed by an objective function. For instance, we can use the distance of the instances to each other or to the centroid. We then try to minimize the squared distance of each instance to its closest centroid. Using the Euclidean distance, we can calculate the sum of squared errors (SSE). In other words, the error is the squared Euclidean distance of each data point to the closest centroid. In two iterations of the k-means algorithm, the set of clusters which produce the smallest total sum of squared errors is preferred. Using the distance measure in equation 6.1, we can calculate the sum of squared errors as:

$$SSE = \sum_{i=1}^{k}\sum_{j=1}^{(n(i))} ||x_j^i - c_i||^2 \qquad (6.2)$$

where $n(i)$ is the number of instances in cluster i, x_j^i is the jth instance of cluster i and c_i is the centroid of cluster i. Steps three and four of algorithm 6.1 attempt to minimize the objective function sum of squared errors. The sum of squared errors is also called the scatter. A smaller sum of squared errors means the centroid, the prototype of a cluster, represents the points in the cluster better than clusters with a higher sum of squared errors. Once the difference of the sum the squared errors between two consecutive iterations of the k-means algorithm is smaller than a specified value, the process stops.

A few remarks: k-means is not restricted to Euclidean spaces. k-means clustering is a very general clustering algorithm that can be used for a very wide variety of data types, including document data and time series. k-means is very sensitive to the initial centroids. It can produce different results on the same data set for different initial centroids. A way to mitigate this problem is to run k-means several times and select the run that produced the best result based on the objective function or the cluster assignment that is mostly observed. Another approach is to first randomly select a sample data set and use hierarchical clustering to determine the number of k.

Since k-means assigns the instances to the closest centroid, the clusters form a hyper-sphere, in other words, they are globular in shape with the centroid being the center. k-means clustering does not work well for clusters that are nonspherical. Pre-processing the data set using transformations can be used to solve this problem.

6.3   Hierarchical clustering

Hierarchical clustering, also called hierarchical cluster analysis (HCA), is a collection of closely-related cluster algorithms that build hierarchies of clusters. Hierarchical clustering does not only partition the data, it also depicts the relationships among the clusters. There are two basic approaches, agglomerative and divisive clustering. Agglomerative clustering follows a bottom-up strategy, where each instance starts as its own cluster. Each data point is fused with one or more other points to form larger clusters based on similarity. In the last step, we end up with one cluster containing all instances. Divisive clustering is a top-down approach and works in the opposite way. The clustering starts with one single cluster that is divided into subclusters. The subclusters are further subdivided in the next iterations until each data point is in its own cluster. Divisive clustering does not need to go through all iterations and can halt once a stop criterion is met. The two approaches are shown in Figure 6.2. Agglomerative approaches tend to be simpler to implement programmatically, whereas divisive approaches seem to be closer to how the human brain works. In practice, agglomerative clustering is more widely used.

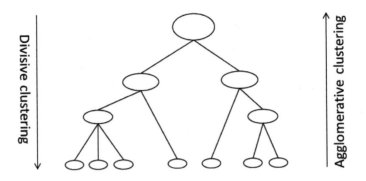

Figure 6.2: Agglomerative and divisive clustering.

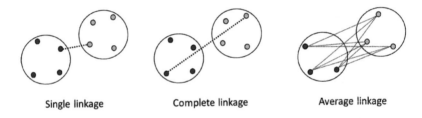

Figure 6.3: Single, complete and average linkage.

Before any clustering can be performed, the distance of each data point has to be determined using a distance function. The distances are stored in a proximity matrix. In each iteration, the proximity matrix has to be updated with the distances between each cluster. The distance between two clusters can be determined by measuring the distance of the two most similar points in each cluster. This is called single linkage or nearest neighbor. In complete linkage or farthest neighbor the two data points that are furthest apart in each cluster, in other words, that have the greatest distance, are measured. In average linkage, the distance between each data point in one cluster with all the data points in the other cluster is measured. The three methods are shown in Figure 6.3.

There is no theoretical justification for the selection of the linkage method. The choice should be made based on the domain of the application. Also, as with virtually any machine learning technique, there are variations and other methods exist. In practice, the Ward method is widely applied. Ward's method uses the increase in the squared error that results when two clusters merge as proximity measure. It uses an objective function approach and an agglomerative hierarchical cluster algorithm. It uses the same objective function that we have seen in equation 6.2 for k-means clustering. In each iteration, the pair of clusters that produce the minimum increase in intracluster variance are merged. Variance is measured by the sum of squared errors. The sum of squares starts out as zero since every data point is in its own cluster. It increases as clusters grow. Ward's method aims to keep the growth as small as possible. To select a new cluster, every combination must be considered, what makes Ward's method computationally expensive. However, it still requires significantly less computation than other methods.

The basic agglomerative hierarchical clustering algorithm is described in algorithm 6.2

Algorithm 6.2 Agglomerative hierarchical clustering algorithm

---

1: *Calculate the proximity matric*
2: *repeat*
3:    *Merge the two closest clusters*
4:    *Calculate the proximities of the new clusters and update the proximity matrix*
5: *until only one cluster remains*

6.4 Visualizing clusters

A hierarchical cluster is often displayed graphically as a dendrogram, a tree-like diagram, as shown in Figure 6.4. A dendrogram shows the clusters, their relationships and the order in which they were fused or split.

A dendrogram can be cut at a certain level to create individual clusters. However, the dendrogram cannot be used to determine the number of clusters. It can often suggest a correct number. In the library example,

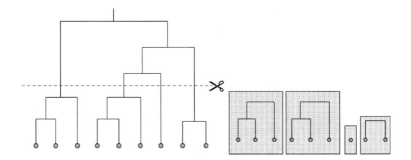

Figure 6.4: Dendrogram clustering.

the user has to define where to make the cut. It might be enough to have one cluster with history books. If there are too many history books it might make sense to use subclusters such as "Ancient world", "Middle ages" and "Modern age" and put them in different shelves.

6.5 Evaluation of clusters

As mentioned before, evaluating a cluster is challenging since there is normally no ground truth is available. Cluster analysis is often used in an exploratory data analysis project. As such, it does not seem obvious why cluster evaluation is important. Evaluating a cluster is important for several reasons:

1. Different cluster analysis methods might find different clusters
2. The correct number of clusters might only become obvious during cluster evaluation
3. Clusters that have been found are not relevant or clusters are found even though there is no cluster data present

If clusters have been found, i.e., there are non-random structures present, we need to determine how accurately the clustering has been performed. If no ground truth, such as labeled data, is available, intrinsic methods have to be used. Intrinsic methods evaluate a clustering by examining how well the clusters are separated and how compact the clusters are [10]. Also, different types of cluster analysis methods require different evaluation methods. For instance, sum of squared errors works well for k-means clustering but will not work well for density-based clusters. The quality of many cluster types can be evaluated using measures such as cohesiveness (compactness, tightness) and separateness (isolation). Cohesiveness measures how close instances in a cluster are to each other. In statistical terms, this is equivalent to having a small standard deviation (i.e., being close to the mean value) [43]. In other words, cohesiveness measures how close instances are to the centroid. Separateness measures how distinct or well separated a cluster is from others. In general, we want clusters that are both, cohesive and well separated. A method that combines both is the Silhouette index.

6.5.1 Silhouette coefficient

The silhouette coefficient is a measure for how similar instances are within a cluster (cohesion), compared to other clusters (separation). The silhouette coefficient ranges between -1 and 1. The higher the index, the more similar an instance is to objects in its own cluster and dissimilar to objects in other clusters. If most instances have a high index, this means the clustering is cohesive and well separated. The silhouette index can be calculated using a distance measure such as the Euclidean distance or the Manhattan distance. The silhouette index is only useful for partitional, not for hierarchical clustering.

To calculate the silhouette coefficient we need to compute $a(o)$, the average distance between o and the other instances in the cluster to which o belongs. $a(o)$ is a measure for the compactness of the cluster. For a data set D with n objects and k clusters, $C_1, ... C_k$, we calculate for each object $o \in D$ where $o \in C_i (1 \leq i \leq k)$:

$$a(o) = \frac{\sum_{o' \in C_i, o \neq o'} dist(o, o')}{|C_i| - 1}$$

(6.3)

and $b(o)$, the minimum average distance from o to all the clusters to which o does not belong:

$$b(o) = \min_{C_i:1\leq j\leq k, j\neq i} \left\{ \frac{\sum_{o'\in C_j} dist(o,o')}{|C_j|} \right\} \qquad (6.4)$$

The silhouette coefficient o is then defined as:

$$s(o) = \frac{b(o) - a(o)}{max\{a(o), b(o)\}} \qquad (6.5)$$

A small $a(o)$ means the cluster is cohesive, a large $b(o)$ means o is well separated from the other clusters. If the silhouette coefficient $s(o)$ approaches 1, the quality of the clustering is greater. If $s(o)$ is negative, o is closer to instances in another cluster. To measure the overall quality of a clustering, the average silhouette coefficient of all data objects o can be calculated. To measure the quality of a single cluster, we can calculate the average $s(o)$ of all the instances within this cluster. If some clusters have much narrower silhouettes from others, this typically means a poor choice of k. The silhouette coefficient can also be plotted in order to heuristically determine the natural number of clusters.

As usual, there are other intrinsic cluster evaluation methods. The Dunn index and Davies-Bouldin index both belong to this class of cluster evaluation algorithms. The Dunn index, like the Silhouette index, is a method for cluster evaluation that depends solely on the data itself, not on any external ground truth. The aim is the same as with the Silhouette index: To obtain clusters that are well separated and compact. The instances within a cluster should have little variance. The higher the Dunn index, the better the clustering. The Silhouette and Dunn index are internal measures since they only use information present in the data set. Cluster evaluation against external data such as a gold standard data is also possible. An example is entropy, which measures how well the clustering data matches external data.

Chapter 7

Semi-supervised Learning

CONTENTS

Supervised methods use labeled data for training, unsupervised methods are applied on unlabeled data and have to determine feature importance by themselves. Semi-supervised methods follow a hybrid approach. Sometimes, we have a small number of labeled instances and a large number of unlabeled instances. For instance, a fraud detection application in a financial institution might be able to detect known types of fraud, but there are many new, unknown types of fraud that the system might miss. In this case, semi-supervised methods can be applied. This has several advantages. Labeling data, which is a time consuming and costly approach, is often performed by a data scientist. Manual labeling can also introduce human bias. Using unlabeled data during training can improve accuracy and often semi-supervised methods perform considerably better than unsupervised methods. For instance, the human genome consists of approximately 3 billion base pairs. Labeling the whole DNA for whole genome sequencing is virtually impossible. Using semi-supervised methods gives access to large amounts of unlabeled data where assigning supervision information would be impossible. Semi-supervised methods attempt to improve classification accuracy by using both, labeled and unlabeled data instead of training the model just on a small portion of manually labeled data or using an unsupervised approach on the unlabeled data. However, using semi-supervised methods is not always possible and we often do not know the distribution of the unlabeled data.

There are two basic semi-supervised approaches: Transductive and inductive learning. In transductive learning, we try to use the labeled data in order to infer the labels of the unlabeled data. This is in contrast to inductive learning, where the goal is to output a prediction function which is defined on the entire space X [4]. In inductive learning, we try to deduce rules from the labeled data that can then be applied to the unlabeled data set.

A transductive approach goes through following steps:

First we train a learner, such as naïve Bayes classifier, to learn the classes. We then apply the trained learner to assign class probabilities to the unlabeled data. This step is called the "expectation" step. We then train a new learner using the labels of all the data. This step is called the "maximization" step. The steps are repeated until the model converges, i.e., the model does not produce a different estimate anymore. This procedure is called the expectation maximization algorithm or EM algorithm. Each expectation maximization iteration generalizes the model more. The expectation maximization procedure guarantees finding model parameters that have equal or greater likelihood at each iteration [40].

7.1 Expectation maximization

Expectation maximization (EM), also called expectation maximization cluster analysis, is a method to solve maximum likelihood estimation problems. Generally speaking, expectation maximization is an algorithm for maximizing a likelihood function when some of the variables in the model cannot be directly observed, i.e., latent variables. Expectation maximization assumes that the data is composed of multiple multivariate normal distributions, which is a strong assumption. It iteratively tries to find an optimal model by alternatingly improving the model and the object assignment to the model. Looking at Figure 7.1, the upper graphic shows black and grey dots. To characterize each group, we can use the mean of the black dots, which is around 3, and the gray dots, which is around 6. This is called the maximum likelihood estimation. Given a set of data, we try to find the value or the values that best explain each data group. If we have the same data points but we do not know which group each instance belongs to, under certain conditions we might still be able to determine the means for each group by iteratively applying expectation maximization. We can use expectation maximization to determine the mean and standard deviation parameters for each group.

The first step is to estimate the mean and standard deviation for each group. These are random values and do not produce good estimates. In Figure 7.2, we have two normal distribution curves. The mean for the first group is around 3, for the second group around 6. In Figure 7.2, the mean of the first bell curve is around 2.5, the mean of the second bell curve is around 7.5. In the second step we calculate the likelihood of each data point by using the probability density function for a normal distribution using the current guesses at the mean and standard deviation for both groups. This tells us for each data point the likelihood that it belongs to one or the other group. These two likelihood values, the likelihood that a data point belongs to one group and the likelihood that it belongs to the other group, are turned into weights in the third

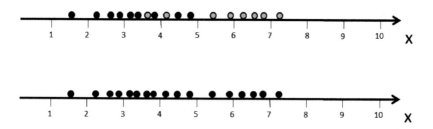

Figure 7.1: Maximum likelihood estimation.

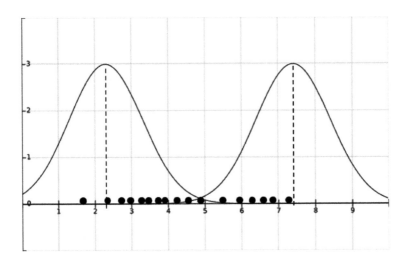

Figure 7.2: Maximum likelihood initial estimation.

step. Using the estimates and the newly calculated weights, we compute new estimates for the mean and standard deviation of both groups. This is step four. The mean and standard deviation are calculated twice using all data points, once for one group and once for the other group. This is the key step where the greater the weight of a data point for one group, the more it influences the estimate for that group in the next estimate. The data point is "pulled" in the right direction, so to speak. Figure 7.3 shows an improved approximation. After we calculate the new estimates, we go back to step two and repeat the process to improve the estimate. We repeat these steps until the estimates converge.

So how do we determine the number of clusters? In cluster analysis, there is no correct solution. Expectation maximization cluster analysis, like all cluster analysis methods, is a knowledge discovery process. The number of clusters selected should be larger than the number of classes. Lets assume we want to train a classifier to tell apart birds, mammals and fish. We cannot assume the data consists of three exact normal distributions. However, we can assume there is more than one type of bird, mammal and fish. So instead of just using three clusters, we create separate clusters for flying birds, flightless birds, etc. So, we might end up with 20 clusters. Once the learner has been trained to

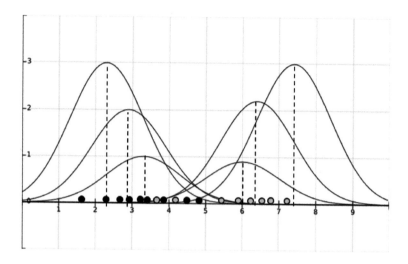

Figure 7.3: Maximum likelihood approximation.

distinguish the 20 classes, we merge the clusters back into the three original clusters. Some might be difficult to classify, such as whales or manatees that live in the sea, but are still mammals. To verify whether there is some structure in the class and not just random data, the algorithm can be applied to single classes.

The expectation maximization algorithm tends to be very slow. Especially with high dimensional data, the expectation step can be very slow. Also, the algorithm can get stuck in a local maxima that is far from the global maxima. We then need to restart from scratch.

7.2   Pseudo labeling

Pseudo labeling is a simple and efficient technique used for semi-supervised learning. It is also used in deep learning. In fact, pseudo labeling can be used for most neural networks and training methods. In semi-supervised learning, we learn the features from the labeled data. We would like to take advantage of the information in the unlabeled data to get a better understanding of the structure of the data. Pseudo labeling can be used to learn from the unlabeled data. The basic idea of pseudo labeling is to use the labeled data set to train a model, then use that model to label the unlabeled data. Pseudo labeling goes through following steps:

1. Train a model or several models using the labeled data set. The training data set might have to be manually labeled.
2. The model which performed best is then used on the unlabeled data to predict the class.
3. Combine the training set with the labels and the one with the pseudo labels.
4. Train the model like before but with the combined data sets.

The pseudo labels might not be correct but they should have a high predicted probability. If the results are not satisfactory, another iteration might be necessary and we have to start from scratch using different methods.

As usual, we divide the combined data set into training and testing data in order to train a learner. There might not be an underlying structure and, as with a clustering problem, there might not be a useful result. Web page classification is a typical problem for semi-supervised learning. We want to classify web pages into "news pages", "discussion forums", etc. There is an abundance of web pages on the Internet, but it is expensive to manually label thousands of web pages. We can use pseudo labeling on raw web pages by training a classifier on a small number of labeled web pages. Then, we can use the trained learner to label the web pages we want to classify as described before. Deep learners often use unsupervised methods for pre-training. The initial weights of the deep neural network are initialized by applying layer-wise unsupervised training. After the weights are initialized, they are fine-tuned using labeled data and the backpropagation algorithm in a supervised fashion. This also works using semi-supervised methods. We can use extra unlabeled data for unsupervised pre-training. Ultimately, we can use semi-supervised methods on any supervised task if we want to enrich our labeled data with unlabeled data. In many cases, even when using older approaches, such as the naïve Bayes classifier, we can obtain superior performance by adding unlabeled data and using semi-supervised learning.

DEEP
LEARNING

Chapter 8

Deep Learning

CONTENTS

Deep learning, a technique with its foundation in artificial neural networks, has emerged in recent years as a powerful tool for machine learning, promising to reshape the future of artificial intelligence [33]. Deep learners rely on large labeled training data sets in order to obtain state-of-the-art results. Deep learning has been applied for image classification, machine translation or speech recognition and deep learners are present in many modern smart phones for voice commands or for face recognition. They are also used for autonomously driving cars, handwriting recognition, drone reconnaissance or coloring black and white images and films, to name a few applications. Despite their success in many fields, they do not work fundamentally different from traditional artificial neural networks. Contrary to older networks, such as the multilayer perceptron, they often use rectifiers as activation function, not sigmoid curves, but they are trained using backpropagation and gradient descent, i.e., stochastic gradient descent. However, due to their massive size, the computations can be very resource intensive since we have to deal with high dimensionality and a large number of features. For instance, the German language has an estimated 330,000 words. In speech recognition, every word is a feature. If we want to recognize animals in images, the network has to be trained with images of all the estimated 8.7 million species that exist in nature from different angles, at different ages and with different sexes. The availability of training data is often a challenge and it can be very costly to obtain labeled training data sets in the required formats.

Many different deep learning architectures are in use today. A number of theoretical ideas had been proposed prior to GPUs (Graphics processing unit), but GPUs have provided the processing power to accelerate the development of deep learning architectures. The calculations of deep learners, such as convolutional neural networks (CNN), can be heavily parallelized and algebraic operations such as matrix products and convolutions can be transferred to GPUs. GPUs have provided the computing power that made many deep learners require to work efficiently. Now there are chips specifically developed for AI operations.

8.1   Deep learning basics

There is no agreed upon definition in literature for when a learner is called a deep learner and when it is a shallow learner. Deep learners are, in essence, artificial neural networks with many layers. Often, neural networks having more than one hidden layer are considered deep learners. Deep learners can be used for supervised, semi-supervised and unsupervised learning. Deep learners have become very popular over the past years due to their impact on computer vision, natural language processing and bioinformatics. A key capability of deep learners is their ability to automatically extract features. Low-level features are abstracted into higher-level features by passing through the different layers of a deep neural network.

Many different deep learning architectures have been proposed. Popular deep learning architectures include convolutional neural networks, deep belief networks and deep recurrent networks.

Deep learners belong to the class of representation learners. Representation learners create an internal representation of the features prior to making the final prediction. Deep learners transform features into multiple representations prior to mapping them to the final prediction. For instance, convolutional neural networks that have been used for object recognition, learn object representations without the need to design features manually. They are inspired by the animal visual cortex and create inner representations of an input image. As the signal passes through the network layers, low-level features are transformed into higher level abstractions by exploiting the hierarchical nature of objects; edges form contours, contours form shapes, shapes form objects, etc.

There are several considerations when using deep learners. Training a deep learner requires an extensive amount of labeled training data. Finding an optimal deep learning architecture can be a laborious task. Overfitting and convergence can cause problems, a problem that is aggravated the more layers the network has. In deep learners, during training, neurons can "die", meaning, they always output the same value for any input. Dead units are also unlikely to recover and will not take a role in the trained network. This mainly happens when rectifiers are used as activation function. Another issue is the vanishing gradient problem. During training, when using backpropagation, the derivative of the error function can become small, causing the weights to not change the anymore. In the worst case, the network stops learning at all. These and some other issues require supplementary learning strategies for deep learning.

8.1.1 *Activation functions*

As mentioned before, deep learners are not very different from shallow neural networks. However, deep learners often use a different type of activation function. The most common activation function used in deep neural networks is the rectified linear unit, or ReLU. It is a very simple function that returns 0 for negative input, and a value for any positive *x*. It can be written as:

$$f(x) = max(0, x) \qquad (8.1)$$

A ReLU is a ramp function and graphically it looks as shown in Figure 8.1. Despite its simplicity, it can handle non-linear and inter-action effects surprisingly well. Non-linear effects simply means that the relationship between the independent variables and the predictions is not linear, in other words, the slope of the graph is not constant. In-teraction effects means that a variable *a* affects a prediction differently depending on variable *b*. For instance, if we want to predict the risk of cardiovascular problems in patients based on body weight, we also need to know the height of the patients.

The reason the ReLU works so well for non-linearities is that ReLUs can be combined in many different ways. Deep learners are, in essence, very large functions. For one, there is the bias for every node that we have seen in Chapters 2 and 4. As we have seen, the bias is just a

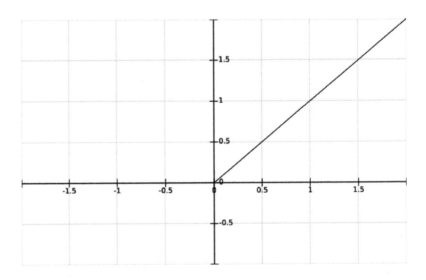

Figure 8.1: Rectified linear unit.

constant that is being learned during training and that shifts the activation function to the left or right along the x axis. Since the ReLU only has two slopes, it is a non-linear function and the bias moves the point where the slope changes. Every node can have its own bias even within a layer and each node can change slope at different points. The second reason seems counter-intuitive at first. As we have seen, traditional networks use sigmoid activation functions and one of the most popular activation functions for shallow neural networks is the hyperbolic tangent, given in equation 8.2. One of the main purposes of the activation function is to account for non-linearities and, since the hyperbolic tangent is non-linear, in fact it is curved everywhere, we might expect that it works well. However, if there are many layers, it causes many difficulties. It is rather flat, except for the small range between about -2 and 2, meaning the derivative is very small outside that range. As mentioned before, deep learners are trained using gradient descent, and since the derivative of the hyperbolic tangent is mostly small, it is difficult to improve the weights. This problem gets worse the more layers the model has. This is called the problem of the vanishing gradient.

$$f(x) = tanh(x) \qquad (8.2)$$

The ReLU has a derivative of either 0 for the flat part of the curve, or 1 for the skewed part. For a large training set, there will usually be instances that yield positive outputs at any given node. So, the average derivative is rarely close to 0. This allows gradient descent to keep improving the weights. However, ReLUs can sometimes be pushed into a state where they become inactive for all inputs and no gradients flow back to the neuron which is a form of the vanishing grade problem. There are variants to the ReLU. For instance, the softplus function is a smooth approximation of the ReLU, as shown in Figure 8.2. It is defined by equation 8.3.

$$f(x) = log(1 + e^x) \qquad (8.3)$$

The derivative of the softplus function is the logistic function, which is also an approximation of the derivative of the ReLU.

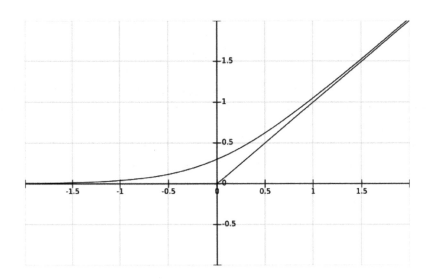

Figure 8.2: Softplus function.

ReLUs have several advantages. They are one sided and have no negative outputs, which is biologically more plausible. They are also efficient for computation since they only use addition, multiplication and comparison operations. Rectifiers have become the activation functions of choice for deep learners since, in comparison to sigmoid functions they allow for faster and more efficient training even on large networks with large data sets. Also, they do not need any semi-supervised pre-processing steps. Even though they can take advantage of semi-supervised setups with extra-unlabeled data, deep rectifier networks can reach their best performance without requiring any unsupervised pre-training on purely supervised tasks with large labeled datasets [7].

8.1.2 Feature learning

Feature learning is a set of techniques that allows a machine to automatically detect features without the need of a feature engineer to manually extract the features of interest. For instance, traditional approaches for object recognition in an image require a feature engineer to isolate the area in the image where the object is located and to omit the background. Using deep architectures, the object of interest is automatically detected, a process called object detection. A deep architecture is capable of doing both, object detection and object recognition. The layered architecture of a deep learner creates higher abstractions of the input in every layer. The output of every layer is a lower dimensional projection of the input. Provided the deep network is optimally weighted, the output of the network is a high-level abstraction of the raw data representing an automatic feature set. This feature set can then be used as input for the actual classification task. Automatic feature detection also helps to avoid human bias. Using, for instance, a convolutional neural network, the convolutional layer acts as a feature extractor.

Feature learning is part of an effort to automate the whole end-to-end machine learning process, called automated machine learning. Automated machine learning aims to automate all machine learning tasks that are usually done manually, such as model selection or hyperparameter. However, automatic feature extraction is not always feasible. Feature learning is not restricted to deep learning and has been used with shallow learners, such as k-means clustering or principal component analysis.

8.2 Convolutional neural networks

Convolutional neural networks (CNN), also called ConvNets, have become very popular for image analysis tasks, such as image classification and tagging, object recognition, face detection and recognition and optical character recognition. They are inspired by the human visual cortex and how it assimilates visual information and have been developed specifically for image analysis. One of the first convolutional neural networks was proposed by Yan LeCun in 1998 [19], one of the three fathers of deep learning, Yoshua Bengio, Geoffrey Hinton, and Yann LeCun [28]. However, the first convolutional network goes further back to the 1980s and many different convolutional neural network architectures have been proposed since then.

Since convolutional neural networks use images as input, some of the image processing properties have been reflected in their architecture. Convolutional neural networks consist of an interleaved set of feed-forward layers implementing convolutional filters followed by reduction, rectification or pooling layers. Contrary to the multilayer perceptron that we have seen in Chapter 4, a convolutional neural network is not fully connected. Also, in a convolutional neural network, neurons are arranged in three dimensions and they accept a three dimensional fixed size input image. The output is a single vector. Otherwise, they are not very different from traditional neural networks that we have already seen and many training techniques that we have seen can be used for convolutional neural networks. The convolutional networks used for object recognition from photos are a specialized kind of feed-forward network [8]. They consist of neurons with learnable weights and biases. They are trained using a loss function, backpropagation and gradient descent. In practice, having more than three layers rarely helps improve the accuracy of a traditional neural network. In contrast, using deep learners, depth is extremely important for a recognition application and a convolutional neural network typically has ten to twenty learnable layers. Since every layer projects a higher level of abstraction of the image, having several layers is intuitive since objects are made up of shapes, shapes are made up of edges, and so on.

In essence, a convolutional neural network consists of five types of layers, an input layer and an output layer and three types of intermediate layers: convolutional layers, pooling layers and non-linearity layers, typically a ReLU.

- ■ The input layer accepts an image, a fixed size three dimensional pixel array: [width x hight x color channel]
- ■ The convolutional layer extracts spatial features from small regions of the image and connects these regions with one unit in the next layer
- ■ The pooling layer reduces neighboring features into single units, this downsampling operation reduces the size of the image
- ■ The ReLU layer applies the activation function, such as $max(0, x)$, the image size remains unchanged

■ The output layer in form of a fully connected multilayer perceptron (MLP) calculates the class score and outputs a vector holding the score

It should be noted that some convolutional neural networks can also process variable sized images.

Figure 8.3 shows an example architecture of a convolutional neural network with a succession of the different types of processing layers. It takes a raw image as input and passes it through the different layers. The output of the fully connected multilayer perceptron is a vector holding the class scores. For ease of representation, Figure 8.3 is drawn in 2D. In essence, the convolutional neural network acts as a feature extractor for the multilayer perceptron. The convolutional layer extracts the relevant features, the pooling layers makes the features equivariant to scale and translation.

The convolutional layer does most of the computational work. The convolutions with learnable parameters are used to extract similar features at multiple locations with few parameters. Receptive fields slide, more precisely, convolve, over the input image, and connect the receptive fields with one unit in the next layer, creating a feature map. The result is an activation map. The network learns feature detectors or filters that activate when they come across a visual feature, such as an edge or a shape of some color. The feature detectors replicate each other and, hence, share the same weights. In this way, a visual feature can be detected anywhere in an image. We end up having a set of

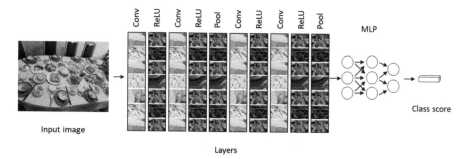

Figure 8.3: Convolutional neural network layers.

convolutional layers with a two-dimensional activation map in each layer.

The pooling layer reduces neighboring features into single units. Pooling is a sample-based discretization process and is also called max pooling since it applies a max filter to a region. The max filter is usually applied to non-overlapping subregions, where each element of the target matrix is the maximum of a region in the original input. However, there are also other types of pooling. The pooling layers are placed periodically in between a series of convolutional layers to reduces the size of the representation and the parameters. This downsampling reduces computing power and creates an almost scale invariant representation. Convolutional neural networks are, thus, able to detect features regardless of where an object appears in an image (location invariance). The pooling layer also helps to avoid overfitting. The whole process is shown in Figure 8.4.

As mentioned before, the ReLU has become the activation function of choice for deep learners since it requires much less computing power and decreases training time. ReLUs are several times faster than, for instance, the conventional tanh function.

The convolutional layer is only connected to a local region of the input. It is not fully connected like the layers in a multilayer perceptron. Also, many neurons in a convolutional layer share parameters. Otherwise, it is not different from a fully connected layer. A lot of the success of

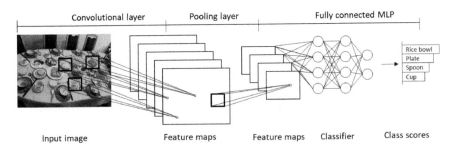

Figure 8.4: Convolutional neural network.

convolutional neural networks lays in their architecture. Many different architectures have been proposed but they are, in essence, variations of the basic convolutional neural network in Figure 8.4. They all use convolutional, pooling and fully connected layers.

Convolutional neural networks are mostly credited for their successful application to computer vision problems. However, they are also used for natural language processing, for video analysis and for medical analysis. They have also been trained to play games, such as checkers and Go. Despite the attractive qualities of convolutional neural networks, and despite the relative efficiency of their local architecture, they have still been prohibitively expensive to apply in large scale to high-resolution images [16].

8.3   Recurrent neural networks

A limitation of multilayer perceptrons and convolutional neural networks is that they require a fixed-sized vector as input and create a fixed-sized output vector. Also, the inputs to traditional neural networks are independent. Recurrent neural networks (rNN) operate over sequences of values $x_1, ..., x_\tau$ and they can produce sequences of vectors as output. The capability of processing temporal sequences makes them ideal for tasks where the input size is not known beforehand. For instance, in machine translation, the length of a text or speech we want to translate from one language into another might not be known beforehand. Consequently, we also do not know the length of the translated text beforehand. The capability of processing sequences of inputs makes recurrent neural networks suitable for tasks such as time series analysis, understanding spoken words or sentiment analysis with variable sized inputs. Most recurrent networks can also process sequences of variable length [8].

The foundational work on recurrent neural networks was performed by John Hopfiled and published in 1982 [13]. Since then, many types of recurrent neural networks have been developed, including Elman networks, gated recurrent units, bidirectional recurrent neural networks and deep recurrent neural networks, to name a few. Contrary to feedforward networks, recurrent neural networks contain directed cycles. However, recurrent neural networks as well as convolutional neural networks are rarely used alone but are often part of a broader architecture often in combination with a multilayer perceptron, as we have seen in section 8.2.

We have already introduced recurrent neural networks in Chapter 4 where shallow learners were covered. In essence, as mentioned before a recurrent neural network adds loops to the network architecture. A neuron in a recurrent neural network might transmit a signal sideways to another neuron within the same layer or it might have a connection with itself in addition to the connection with a neuron in the next layer. The output sequence of the network might also be used as feedback with the next input sequence. Figure 8.5 shows some possible recurrent network architectures.

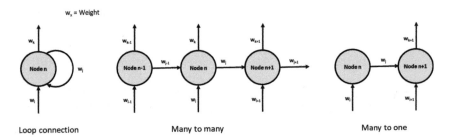

Figure 8.5: Recurrent network types.

The recurrent connections add memory or state to the network. The state is usually stored in form of a state vector that is under the networks control. The state vector is combined with the input using a learned function to produce a new state vector. In this way, a recurrent neural network does not only take into account the current input but also what it has perceived back in time. The decision at time step t is influenced by time step $t-1$. This is very intuitive since we decide based on past experiences. The sequential information is preserved in a hidden state that allows the recurrent neural network to find correlations over time, called long-term dependencies. The hidden state h at time t can be calculated using equation 8.4:

$$h_t = \sigma(Wx_t + Uh_{t-1}) \tag{8.4}$$

where

h_t	=	Hidden state at time step t
x_t	=	Input vector at time step t
W	=	Weight matrix
U	=	Transition matrix
h_{t-1}	=	Hidden state at time step $t-1$
σ	=	Activation function

The transition matrix is the hidden-state-to-hidden-state matrix and represents the memory of the recurrent neural network. The hidden state contains traces of all previous hidden states, not just the last one. The weight matrix W determines how much importance to accord to the input in the same way as we have seen with other types of networks. The activation function σ is usually a logistic or a hyperbolic tangent function.

A recurrent neural network is not necessarily a deep learner. As shown in Figure 8.5, a recurrent network is also organized into layers and a recurrent neural network is usually called deep recurrent neural network if it has more than one hidden layer.

A recurrent neural network is trained using backpropagation and gradient descent, just like a feedforward network. However, because of the loops, the backpropagation mechanism does not work in the same way like for a feedforward network. In a feedforward network, the algorithm moves backwards from the final error through the layers and adjusts the weights up or down, whichever reduces the error. To train a recurrent network, an extention of backpropagation called backpropagation through time, or BPTT is used. Backpropagation through time is a gradient-based method for training recurrent networks. In backpropagation through time, time is defined by the ordered series of calculations moving from one time step to the other. In essence, the structure of the recurrent neural network is unfolded. A copy of the neurons that contain loops is created and the cyclic graphs of the recurrent neural network are transformed into acyclic graphs, turning the recurrent neu-

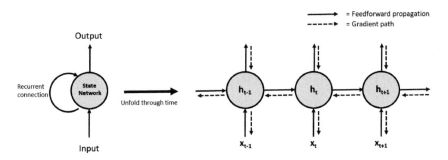

Figure 8.6: Unfold through time.

ral network into a feedforward network. Every copy shares the same parameters. A recurrent network and the unfolded network are shown in Figure 8.6.

The role of backpropagation through time in explaining learning through time in the brain, however, has particular problems not faced by feedforward backprop, and its relationship to the brain is less well-studied in general [21]. Neural networks, whether recurrent or not, are nested functions of the form $f(g(h(x)))$. In an recurrent neural network, the loops extend the series of functions for which we have to calculate the derivatives and apply the chain rule.

Not all recurrent neural networks can be unfolded. With some architectures, it is not possible to use backpropagation through time. Also, as usual, there are variations of backpropagation through time, such as truncated backpropagation through time, which is an approximation of the full backpropagation through time. Truncated backpropagation through time is used when there are many time steps making computation slow. Also, recurrent neural networks, particularly when they have many layers, are plagued by the vanishing gradient problem we have seen before. If we cannot determine the gradient we do not know how to adjust the weights in a direction that decreases the error. In the worst case, the network stops learning at all. Many solutions to the vanishing gradient problem have been proposed. As we have seen before, when using a ReLU as activation function we do not have this problem. Another solution is using residual networks, or ResNet. A residual network is an ensemble of many short, typically three layered networks. ResNets split a deep learner into smaller chunks. ResNets do not require any additional parameters or other training mechanisms. They provide one of the most effective solutions for the vanishing gradient problem.

Recurrent neural networks are very effective for text and speech analysis as well as time series analysis, as we have seen. They are also used for robot control, for text recommendation and text autofill or customer satisfaction analysis in call centers. They work together with ConvNets for image recognition and provide a description to name a few applications.

8.4   Restricted Boltzmann machines

Restricted Boltzmann machines or RBMs are shallow, two layer networks that can be used for classification, regression, dimensionality reduction, collaborative filtering or feature learning. They are less well known than other network types. However, they achieved state of the art results in the Netflix prize [36] and outperformed most of the competition. They are covered in the deep learning chapter because they constitute the building blocks of deep belief networks.

Restricted Boltzmann machines are probabilistic, i.e., they do not assign discrete values but probabilities. A restricted Boltzmann machine is a type of Boltzmann machine. However, Boltzmann machines are recurrent networks, whereas restricted Boltzmann machines do not allow intralayer connections between hidden units. That is why they are called "restricted".

Restricted Boltzmann machines consist of just two layers, an input layer and a hidden layer, as shown in Figure 8.7. They do not have an output layer. The predictions are made differently from regular feedforward networks. They fall into the category of energy-based models. Energy-based models try to minimize an energy function where high energy means bad accuracy. Restricted Boltzmann machines are divided into visible and hidden units. The visible units receive the train-

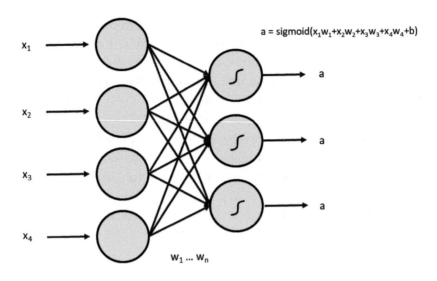

$a = \text{sigmoid}(x_1 w_1 + x_2 w_2 + x_3 w_3 + x_4 w_4 + b)$

Input   Visible layer   Hidden layer

Figure 8.7: Restricted Boltzmann machine.

ing set. The energy function is minimized during training by adjusting the weights, biases and the states of the visible and hidden units until the function reaches a minimum, a process called simulated annealing. Each visible node is connected to every hidden node. A visible node accepts a low level feature, e.g., a pixel from an image. If an image has 1024 pixels, the restricted Boltzmann machine requires 1024 visible nodes. The inputs are combined with weights and added to a bias. The summed up inputs are then passed to a sigmoid activation function. This is not different from what we have seen with other networks. However, restricted Boltzmann machines are not trained using stochastic gradient descent. They use Gibbs sampling and contrastive divergence to update the weights. The training mechanism is beyond the scope of this book.

If there is more than one layer, the output *a* of the hidden layer is used as input for the next hidden layer.

Restricted Boltzmann machines learn how to reconstruct the input in an unsupervised fashion. The activations *a* are used as input in a backwards pass. They are multiplied by the same weights the input x was multiplied. The summed up products are added to a bias of the visible layer. The biases of the visible layer are different from the biases of the hidden layer. The resulting output is a reconstruction or approximation of the original input. During the forward pass, the restricted Boltzmann machine calculates the probability of the output given a weighted x: $p(a|x;w)$. During the backwards pass, the restricted Boltzmann machine estimates the probability of the inputs x given the weighted activations a: $p(x|a;w)$. Both estimations lead to the joint probability distribution of x and a: $p(x,a)$. The training goes through several forward and backward passes. The weights are adjusted until the probability distribution of the original input $p(x)$ and the probability distribution of the estimated input $q(x)$ converge.

Once the restricted Boltzmann machine training is completed, the activations of the hidden layer become the input of the next hidden layer. The second hidden layer is trained until it is able to approximate the input from the previous hidden layer. This step by step approach of training hidden layers in an restricted Boltzmann machine is one of the most common deep learning strategies since it permits to train many hidden layers efficiently. It is unsupervised, since it does not require labeled data for training. This sequential construction of a multilayered network allows to learn more complex representations by creating a feature hierarchy. The pre-trained multilayered restricted Boltzmann machine forms the basis for a deep belief network. The pre-trained deep belief network can then be trained for instance for image classification using supervised learning. Pre-training a network by initializing the weights is one of the main use cases for restricted Boltzmann machines.

8.5   Deep belief networks

A deep belief network (DBN) is a sequence of the restricted Boltzmann machines that were introduced in the previous chapter. It is a type of deep learner that consists of multiple hidden layers with connections from previous and to subsequent layers but not within layers. Like a restricted Boltzmann machine, it can learn to reconstruct its input. The layered architecture allows for fast unsupervised training. The pre-trained deep belief network is a type of unsupervised pre-trained network. Each layer acts as a feature detector. Once training is completed, it can be trained further in a supervised fashion in order to create a classifier. The deep belief network might then end with an output layer, such as a softmax or logit layer, for classification. A deep belief network can be trained in a greedy fashion, where each layer is trained separately. Each hidden layer becomes the visible layer for the next hidden layer.

There are many use cases for a deep belief network. A deep belief network can be used for classification, image recognition or image generation. It can also be used for clustering. In this case there is no output layer. Other use cases of deep belief networks are video sequences and motion-capture data generation. In practice, deep belief networks have many real-world applications. They are used to classify high resolution satellite or medical images, for drug discovery or to analyze electroencephalograms.

A deep belief network accepts binary input as data. An extension, continuous deep belief networks, also accepts continuous decimal values.

8.6 Deep autoencoders

An autoencoder is a network used to learn a representation (encoding) of a data set in an unsupervised manner. An autoencoder compresses the data into a code, then uncompresses it again into something close to the input. In this manner, it is trained to ignore noise in data. The autoencoder learns both to compress, i.e., encode, the input and to reconstruct the input. Autoencoders have gained popularity in the past years since they have been included in many powerful deep learning architectures. They are typically used for dimensionality reduction. Autoencoders can also be used as a generative model. The learned codes can be used to construct an image of something the autoencoder has not seen before.

In its simplest form, an autoencoder is a feedforward network with an input, an output and one hidden layer. The input and output layer have the same number of neurons. An autoencoder always encodes and decodes an input. Autoencoders try to learn a function h such that $h_{W,b}(x) \approx x$ where the parameters W are the weights and b is the bias. In other words, they are trying to approximate the identity function. An identity function always returns the same value that was passed to the function as an argument. Autoencoders are trained in an unsupervised fashion by minimizing a loss function such as the squared error. If the hidden layers are larger than the input layer, an autoencoder can learn the identity function and reconstruct the input identically.

Deep autoencoders consist of two symmetrical deep belief networks. They are capable of compressing data, such as images. They can also be used for topic modeling. Topic modeling is the process of discovering the topic or the topics in a set of documents.

Deep autoencoders consist of one encoding part and one symmetrical decoding part. Each part consists of one or more deep belief networks. Figure 8.8 shows a simple deep autoencoder architecture.

An autoencoder takes binary data as input, such as pixel arrays, but it can also use real-valued data as input. If the input is an image of $32 * 32 = 1024$ pixels, the input layer should contain about 1200 units, slightly larger than the 1024 input pixels due to the limited capacity of

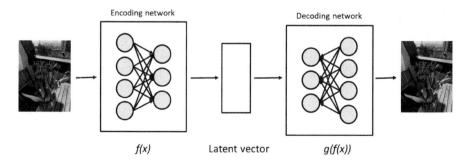

Figure 8.8: Deep autoencoder.

the sigmoid belief units. The subsequent number of units will be something like 600, 300, 150 and 50, where 50 is an encoded version of the input image. It is the latent vector with the compressed representation. Latent variables cannot directly be observed. They are inferred from the observed variables. The decoding half of the autoencoder consists of the same number of layers and nodes but in reverse order. It accepts the condensed vector as input and learns how to reconstruct the image. There are also variations where the hidden vector is bigger than the input layer. If the hidden code is smaller than the dimension of the input, the autoencoder is said to be undercomplete. If it is bigger, it is called overcomplete.

Autoencoders may be thought of as being a special case of feedforward networks and may be trained with all the same techniques, typically minibatch gradient descent following gradients computed by backpropagation [8]. Learning happens through minimizing a loss function L that can be described as shown in equation 8.5:

$$L(x, g(f(x))) \tag{8.5}$$

by penalizing $g(f(x))$ for being dissimilar to x.

Typically, when using an autoencoder, we are not interested in the output but rather in the properties of the compressed representation. The representation can be used for topic modelling, but is also useful for information retrieval in documents since the hidden code contains the most salient features. They are also used for image search engines. When the hidden code of an image is similar to the hidden code of another image, it is translated into the matching image that is then returned. Autoencoders have also been developed using ConvNets as encoding and decoding networks instead of deep belief networks.

LEARNING
TECHNIQUES

Chapter 9

Learning Techniques

CONTENTS

In machine learning, learning is a search through the space of possible representations to find the representation that best fits the data at hand. As such, learning can be considered a search. Machine in machine learning refers to the algorithm that performs the search. An algorithm is a combination of math and logic. Typically the search space is too large to be searched entirely. This is sometimes called the curse of dimensionality. Techniques such as dimensionality reduction can be used to alleviate the problem. There are also learning techniques that help improve the performance of a learning algorithm in high dimension.

9.1 Learning issues

Learning a representation is a compromise between many desiderata [31]. The richer the representation, the more useful it usually is for subsequent problems. However, in order to create richer representations, a lot of data is required. Often, the availability of required data is a problem. Techniques such as cross-validation help to mitigate a data scarcity problem. Also, machine learning schemes are prone to overfitting. "Fit" in a statistical sense means we seek the best description that fits the data reasonably well [40]. An overfitted model does not generalize well from training data to unseen data and captures noise instead of the underlying signal. Basically, the learned model is too closely tied to the particular training data from which it was built [40]. An overfitted model is also called overtrained.

In predictive modeling, the signal is the true underlying pattern that you want to learn from the data. Noise refers to irrelevant information or randomness in a dataset. When the representation cannot adequately capture the underlying structure of the data, it is called underfitted or underlearned. Underfitted models tend to have less variance in their predictions but more bias towards wrong predictions. Bias is the difference between the expected and the measured (observed) prediction. Variance is the sensitivity of the learning algorithm to specific data sets. A model with high variance will produce very different models depending on the training data set used. Learners with low variance and high bias tend to be simpler, but have a rigid underlying structure. Naïve Bayes or regression models fall into this category. On the other hand, learners with low bias and high variance tend to be more complex, but have a more flexible underlying structure. Decision trees with high depth and k-nearest neighbor learners fall into this category. This tradeoff in complexity is called bias-variance tradeoff, since a learner cannot be complex and simple at the same time. Ensemble learners tend to have less bias and variance errors.

9.1.1 Bias-variance tradeoff

The bias-variance tradeoff is a fundamental problem of supervised learning and applies for classification and regression. It is the problem of reducing two sources of error, bias and variance, simultaneously. To find a good balance between bias and variance, a good predictive model reduces the total error. The total error is defined as:

$$Total\ error = Bias^2 + Variance + Irreducible\ error \qquad (9.1)$$

where

Bias	=	Deviation between the expected and the actual value
Variance	=	Variance is error from sensitivity to small fluctuations in the training set
Irreducible	=	Irreducible error is the noise term in the true relationship

Bias refers to model error and variance refers to the consistency in predictive accuracy of models applied to other data sets [1]. The irreducible error, also called noise, cannot fundamentally be reduced during training by any model. It can be reduced sometimes by performing better data pre-processing steps, usually better data cleansing. The irreducible error stems from the data itself, bias and variance from the choice of the learning algorithm. With infinite training data it should be possible to reduce bias and variance towards 0 (asymptotic consistency and asymptotic efficiency). However, some learners are not capable of fitting a model even if enough data is available. For instance, algorithms that linearly separate the data, such as the perceptron, cannot fit data that has a non linear pattern. Figure 9.1 shows data that can and cannot be fitted using a linear algorithm.

Bias and variance are two independent criteria for assessing the quality of a learner. Ideally, we want low bias and low variance. However, this is difficult to achieve simultaneously. Unfortunately, low bias on training data usually means high variance because the model was overfit. Low variance usually comes with the cost of high bias. For instance, when training a k-Nearest Neighbor learner, a small value for k provides the most flexible fit, which will have low bias but high variance [14]. The high variance is due to the fact that a low k means a prediction in a region depends on only one observation. The mean squared error (MSE) loss function that we have seen in Chapter 2 in equation

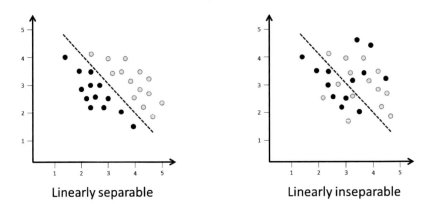

Figure 9.1: Linear fitting.

2.9 combines bias and variance, which makes it a very good evaluation method. The mean squared error is equal to the sum of the variance and the squared bias of the estimator. It can be partitioned in bias and variance, as shown in equation 9.2.

$$MSE = Bias^2 + Variance \qquad (9.2)$$

where

Bias	= Error from wrong assumptions of the model
Variance	= Variance is error from sensitivity to small fluctuations in the training set

Since we cannot reduce bias and variance at the same time, we need to find an optimal balance, as shown in Figure 9.2. There are several strategies to manage bias and variance. Simplifying the model can decrease variance, for instance through feature selection or dimensionality reduction. Also, using a larger training set tends to decrease variance.

As mentioned before, different learners often behave very differently and one key differentiator is how much bias and variance they produce. That is why usually several learners are trained and evaluated to find the best suitable algorithm for the problem. Another approach is us-

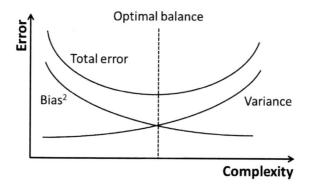

Figure 9.2: The bias-variance tradeoff.

ing ensembling. Bagging is a simple yet powerful ensemble method. It is a general procedure that can also be used to reduce variance. Bagging, short for bootstrap aggregating, is a resampling technique where different training sets are created using random selection with replacement (meaning we can select an instance multiple times). Each sample is used to train a model. The predicted values are then combined through averaging and majority-voting. Although bagging was first developed for decision trees, the idea can be applied to any algorithm that produces predictions with sufficient variation in the predicted values [1]. To get good bagged ensemblers, the individual models are overfit, meaning the bias is low. Bagging is reducing variance by averaging the predictions. Bagging can be considered a variance reduction technique. Another feature of bagging is that the resulting model uses the entire feature space for node splits. The biggest weakness of bagging is that it might produce correlated trees, thereby increasing bias and reducing variance. Random forests, that were discussed in Chapter 4, are a simple yet effective way to avoid inter-tree correlations.

As mentioned, managing bias and variance is essentially dealing with under- and overfitting. Applying an appropriate learning method is crucial for obtaining well fitted models. Which learning method is most suitable depends on several factors. For instance, if there is only limited training data available, cross-validation can be an effective learning method. Cross-validation divides the available data into subsets that are used alternatively for training and testing and, thus, eventually exploiting all data in an optimized way. Dimensionality reduction, which refers to the process of reducing the number of variables, might be used to prevent overfitting. Dimensionality reduction can mean feature selection, basically selecting a subset of the original variables for training, or feature extraction, which means reducing the dimensionality of the space. Both methods can be used to prevent overfitting. It should be noted that it is often computationally not feasible to find the best fit.

9.2 Cross-validation

Cross-validation is a statistical method for assessing how well machine learning models generalize on new data. It is usually used for predictive tasks in order to evaluate how accurately a model will perform in practice. It is often used to compare and select predictive models since it is easy to implement and interpret. The results tend to have lower bias than other models.

A frequently used type of cross-validation is k-fold cross-validation. k-fold cross-validation is a resampling procedure that can be used when there is limited training data. Whereas normally a data set is split into training, validation and test data, k-fold cross-validation uses all observations for training and testing. The basic ideas is to split the training data into k equal sized folds randomly and use $k-1$ folds for training and the remaining hold out fold for testing. This procedure is repeated k times until every fold has been used exactly once for testing. In this way, all the instances are used for training and every observation is used for validation only once. If, for instance, $k = 10$ it is also called 10-fold cross-validation.

The only configurable parameter in k-fold cross-validation is k. Usually, setting $k = 10$ yields a learner that performs well. Extensive tests on numerous different datasets, with different learning techniques, have shown that 10 is about the right number of folds to get the best estimate of error, and there is also some theoretical evidence that backs this up [40]. However, when a cross-validation experiment is repeated, and a new set of folds is randomly selected, the results can often be different. A common approach is to repeat 10-fold cross-validation 10 times and then average the results. The downside is that, in this case, the learning algorithm is invoked 100 times, which can be computationally expensive.

A variation is leave-one-out cross-validation. Leave-one-out cross-validation is an extreme version of k-fold cross-validation where k equals n, the data set. The learner is trained on $n - 1$ observations and a prediction is made for the remaining observation point. As before, this process is repeated until each data point has been used for validation once and the results are averaged. This sounds computationally expensive, but there are implementations that do not require more computing time than other methods.

Stratification is a technique that is often used with cross-validation. Stratification helps reduce variance but it does not eliminate it entirely. When using stratification, the folds are selected in such a way as to assure equal distribution of the class labels in all folds to avoid over- or underrepresentation of a label in any data subset. Stratification was explained in section 3.2. When used with stratification, it is called stratified k-fold cross-validation. When used, for instance, for binary classification, the proportion of the two classes are roughly equally distributed in each fold. Neither the distribution nor the folds have to be exact. The data is divided into approximately equal folds. Stratified 10-fold cross-validation has become the quasi standard method and it is a good starting point if choosing k is causing difficulties. Usually, the results do not differ much if 5 or 20 is set for k.

9.3 Ensemble learning

Ensemble learning, or ensembling, combines several trained models from the same data, usually of the same class. It is often advantageous to take the training data and derive several different training sets from it, learn a model from each, and combine them to produce an ensemble of learned models [40]. Ensembling often allows to improve the predictive performance of a model by combining several rather weak learners to create a strong one. If several learning schemes have been trained, instead of using the best performing one, the results of all the trained learners are combined. Ensembling can be used for supervised and unsupervised learning.

Two common ensemble learning techniques are bagging and boosting. In bagging, an abbreviation of bootstrap aggregating, several similar learners are trained. Another bagging technique trains several different learners using different training data sets. Then the results are combined either through averaging or through modal votes. This helps reduce overfitting and variance. Variance reflects the sensitivity of the model and it increases when the complexity of the model increases. A model with high variance tends to be overfitted and does not generalize well beyond the training data set.

In boosting, a model is incrementally created. Each model is trained on the instances the previous model misclassified. The most popular boosting algorithm is AdaBoost, adaptive boosting. AdaBoost is less prone to overfitting, but it is sensitive to noise in the data or to outliers. AdaBoost is a heuristic approach that potentially improves the performance of the learner. A problem of machine learning is that there are potentially a large number of features, the problem of high dimensionality. AdaBoost only selects the features that potentially improve the predicting power of the model by reducing the dimensionality and, thus, reduces execution time.

9.4   Reinforcement learning

Reinforcement learning is a reward-based learning technique. It is inspired by nature and can be compared with, for instance, training a dog to stand on its back feet. The dog gets a dog biscuit when it learns to stand on the back feet. Similarly, in reinforcement learning, there is a reward function used to train an agent how to act. Reinforcement learning is different from supervised learning since no correct input/output pairs need to be present for training. In essence, in reinforcement learning, an agent learns to react to an environment. It is considered a separate learning paradigm from supervised and unsupervised learning.

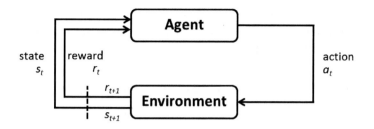

Figure 9.3: Reinforcement learning.

Reinforcement learning is a stepwise approach, as shown in Figure 9.3, where at each time step t the agent can choose an action a from a set of available actions. The agent goes through sequences of state-action pairs until it predicts the best path for the agent to take. The prediction is modelled as a map called a policy. At each time step t, the environment, with which the agent interacts, moves to a new state s at time step $t+1$ and the reward r for the transition to the new state is computed. The reward can be positive or negative. The reward is calculated using a reward function that takes the state of the agent and the action as input and maps it to a possible reward. The goal of the agent is to pick the best possible action for any state and to collect as much reward as possible. The way the actions are selected is crucial for the performance of the learner.

Reinforcement learning can be used with neural networks and has been particularly successful in recent years in combination with deep learners. Deep learners are neural networks that have many hidden layers between the input and output layer. Reinforcement learning in combination with deep learning has been successfully applied to many problems, such as playing games, machine translation and robot control.

9.5   Active learning

Active learning is a form of online learning in which the agent acts to acquire useful examples from which to learn [31]. Active learning is an iterative type of supervised learning that is suitable for situations where data are abundant, yet the class labels are scarce or expensive to obtain [10]. It is a special case of learning in which the learner can actively query the user or some other information sources to obtain labels. The idea behind active learning is that the learning algorithm can perform better if it chooses the data it wants to learn from by itself. Since the learner chooses the examples presented to the user, the number of examples can be considerably lower than the number required for supervised learning. Active learning is an approach that helps identify informative unlabeled examples automatically [23].

In active learning, a small subset of the unlabeled data is randomly selected and labeled manually. Based on the data labeled so far, new instances are presented to the user for labeling. The learner can select superior data points from random sampling. This is repeated until no more useful instances can be found.

The advantage of active learning is that the algorithm can choose the examples to label, and subsequently perform better with a smaller amount of data from methods that randomly select samples from the underlying distribution. How to select the data points to label is a major area in research.

9.6 Machine teaching

Machine teaching happens when a model is not trained by a human, but devices communicate with each other and share knowledge. Basically, they learn from eachother. This can be cooperative, for instance, when autonomously driving cars share information about road conditions, or it can be competitive, for instance two machines playing chess against each other. The machines share data and create synthetic experiences for each other and cause a self-amplifying effect. One big advantage is that machine teaching potentially bypasses the need for training data. The other advantage is that machine teaching helps avoid human bias.

9.7 Automated machine learning

A lot of effort has been made to automate machine learning since many tasks are time consuming and require skilled data scientists that are not easy to find. However, while some machine learning steps can be automated, human experts are still required for many tasks. Many tasks require a trial and error approach since there is no recipe for what method works best.

Tuning the hyperparameters of a learner is something usually done by a data scientist. The goal of hyperparameter tuning is to find the global optimum as fast as possible. The right fine tuning requires experience and different combinations have to be tried and evaluated. Hyperparameter search is something that can be automated and there are already many libraries available for automating the search. TensorFlow's Vizier is a service for black-box optimization of many learners. Other libraries for hyperparameter tuning include Eclipse Arbiter and the Python library Spearmint.

To a limited degree, feature selection can be automated. Some tools can select the features based on their relevance and usefulness for the problem. Automated feature selection is particularly useful in situations where there are too many for a human to choose. However, the most relevant features are not necessarily the best explanatory variables for a specific problem. Even if features are selected automatically, a data scientist will still most likely look at the variables in order to verify that they are relevant. Feature selection and verification requires domain knowledge and features are usually problem dependent and are, thus, selected manually for each data set from scratch.

Model selection involves comparing the prediction accuracy of several trained learners. It is one of the more obvious tasks to automate, since selecting the best performing model involves comparing numerical values. Nevertheless, results might not always be conclusive and different evaluation metrics might be optimal for different models. For instance, one learner might have a good accuracy but a poor F_1-score, whereas for the other it is the opposite, which makes it not always obvious which one to chose.

Often, the results of a machine learning application need to be evaluated by a human expert. Whether the result is useful for the problem at hand or not requires human judgement and cannot be automated. Generally speaking, machine learning requires domain knowledge and an understanding of the data used, which limits the tasks that can be automated.

MACHINE LEARNING APPLICATIONS

Chapter 10

Machine Learning Applications

CONTENTS

Machine learning has applications in virtually every domain. Whether it is in finance, science or academia, as soon as a lot of data needs to be analyzed or many and complex rules are required, machine learning should be considered. It should be noted that, despite the versatility of machine learning, it is not a solution for any type of problem and it is certainly not a silver bullet. If the predictions can be made using simple rules, it is possible to develop a robust solution without machine learning. Machine learning should be considered when:

- Complexity: The rules are too complex or there are too many rules
- Scalability: The data volumes are too large for traditional data mining methods

It is also important to realize that machine learning does not always provide results. Sometimes the data quality is too poor or there is no pattern in the data at all. Before a machine learning project is started, the following conditions should be verified:

- There is an infrastructure to conduct a machine learning project
- There is enough data and the data is of good enough quality
- There is a clear goal for the project

Machine learning works well if there is a lot of data and if there is a ground truth. Recent technological developments, such as the advent of the Internet of Things (IoT), mobile devices and cloud computing, are likely to produce an ever increasing amount of data. Machine learning provides the functionality to effectively analyze the volumes of data produced and it is to be expected that the importance of machine learning will increase in the future. Machine learning technology is already included in many mobile phones and Internet of Things solutions and will seamlessly be integrated in many applications to come. This chapter gives an overview of machine learning applications, but by far not a comprehensive one.

10.1 Anomaly detection

One of the most used applications of machine learning is anomaly detection. Anomaly detection refers to a wide range of unusual activities, such as fraudulent bank transaction, network security breaches or abnormal sensor data suggesting imminent machine failure.

The increasingly digitized world and the advent of Internet of Things produces an ever-increasing amount of transactional and sensor data. The main goal of digitization is to increase productivity and efficiency and ultimately reduce costs. Machine learning is used to identify the data points and patterns that are relevant to optimize capacity, but it is also used to detect observations that are suspicious since they are deviant from the majority of the data. Deviant data is usually caused by some kind of problem, such as bank fraud, malfunctioning equipment, network intrusions or medical problems.

There are different methods to identify abnormal data points. In some simple cases, data visualization can be used, for instance, to detect outliers that are located outside the normal data distribution. However, it is often not one variable alone that allows us to find an anomaly but a combination of many variables or a pattern that need to be identified.

The easiest way to detect anomalies is to verify that certain values, such as sensor data measurements, are out of bounds. If some threshold is exceeded, an alarm is sent. The problem with this approach is that there is the risk of having a high false positive rate or that some anomalies go undetected. Usually, we cannot consider single sensor values on their own when determining the health of an equipment, we need to take into account combinations of sensor values, in other words, patterns.

10.1.1 Security

One problem with anomaly detection is that an anomaly might be unseen and there is no historic data available that reflects the anomaly. For instance, there are constantly new security threads detected that have been unknown and, hence, a machine learning algorithm cannot be trained on existing data. A zero-day vulnerability is such an unknown or unaddressed security vulnerability that can be exploited by hackers. Since there is no data yet from unknown vulnerabilities or security breaches, instead of training a machine learning scheme to detect anomalies in network traffic, the learner can be trained to recognize what normal network traffic looks like. If it detects some deviant traffic, it issues an alarm and a security specialist can analyze the suspicious traffic. What deviant traffic is depends on what is transmitted over the network. For instance, if the network is used for financial transactions, an anomaly might be unusually high amounts of money transferred or transfers to an unusual country. When we have a collection of data points from network traffic they typically have a certain distribution, such as a Gaussian distribution. To detect anomalies, we first calculate the probability distribution $p(x)$ from the data points. For every new data point x, the probability that it belongs to the probability distribution is calculated and compared against a threshold. If $p(x)$ is smaller than the threshold, x is considered an anomaly. Normal data points tend to have a large $p(x)$, whereas anomalous data points tend to have a small one.

10.1.2 Predictive maintenance

Predictive maintenance is used to monitor the health state of an equipment. It helps to avoid costly outages of production sites by determining if a part or a whole machine is going to fail soon and needs maintenance. Modern machines are equipped with sensors that collect data, such as the temperature, humidity, vibration or other values, depending on the type of equipment. By analyzing the sensor values, problems such as material fatigue or abrasion can be determined and the affected parts can be replaced before the whole machine or even a whole production site fails. For instance, airline companies cannot have a spare parts depot in every airport they land, especially for expensive parts, such as propulsion engines that cost several millions. They need to know in advance if an airplane needs maintenance. Modern airliners are equipped with thousands of sensors in the wings and engines for early detection of technical problems. Replacement parts can then be ordered when needed and no warehouse needs to be maintained. This is called smart logistics since it optimizes the logistics of spare parts and the scheduling of maintenance personnel. The service technicians can be scheduled in advance and deployed when needed. The early detection and avoidance of technical problems is also called predictive and preventive maintenance. Smart logistics and preventive maintenance are two main application fields and drivers of Internet of Things.

One solution for predictive maintenance is to use the approach described in section 10.1.1 and determine if a data point belongs to a probability distribution or not. Another approach uses an autoencoder. Autoencoders were described in section 8.6. The autoencoder creates a lower dimensional representation of the sensor values with their correlations and interactions and reconstructs them back to the original values. The autoencoder is trained on normal sensor data that represents the normal operating state. If the monitored equipment degrades, the interactions between the variables change, which results in an increase of the error of the reconstructed input values. We use the probability distribution of the reconstruction error in order to determine if a data point is normal or if we have an anomaly.

10.2 Biomedicale applications

10.2.1 Medical applications

Machine learning has found a large application field in pharma and medicine. It is used to support physicians in disease identification and diagnosis, for medicine discovery to find new drugs or for personalized treatments.

Machine learning methods and, in particular, convolutional neural networks (CNN) that were described in section 8.2 have been successfully applied to analyze medical images such as CAT scans, X-rays and MRI scans. Analyzing medical images is time-consuming and there is always the danger of overlooking a medical problem. A convolutional neural network can detect pathogenic abnormalities in images that are hidden to the human eye. Convolutional neural networks have been used for early detection of illnesses such as Alzheimer's or fibrosis. They are also used to classify images or tissue samples into malign or benign tumors.

Machine learning also has great potential for remote monitoring and diagnoses. A patient can take a picture of a wound or of visible symptoms and upload it to a server in a medical facility where a doctor can look at it and use machine learning for diagnosis and prescriptions. Remote monitoring is not limited to images. Data from a health tracker, such as blood pressure or blood sugar levels, can also be transmitted. Machine learning can significantly improve healthcare by allowing patients who need to regularly see a doctor or who need stationary treatment to be monitored remotely, allowing them to live a more independent life. Since all industrialized countries have a rapidly aging population and ever-growing health care costs, remote monitoring and treatment is a main focus of machine learning in eHealth.

Personalized treatment remains, in many cases, restricted to people who can afford it. Based on the genetic material and on the symptoms, predictive analysis can help provide personalized treatment on scale. For instance, chemotherapy that is not working anymore because the tumor has adapted needs to be adapted accordingly. Machine learning is a major enabler of personalized treatments since it has the capability to analyze large amounts of data, but also to correlate different types of data such as leukocyte levels and genetic information. Using predictive analysis for disease assessment and management helps physicians to select the right treatment and adapt it quickly if some values have changed.

10.3 Natural language processing

Understanding human language remains one of the major challenges in computer science. Natural language is the language humans speak and write and the attempt to extract information from texts or speeches using algorithms is called natural language processing or NLP. Natural language processing encompasses understanding human language, e.g., understanding voice commands on a Smart Phone, and generating natural language, e.g., a text to speech application that reads Web pages to people with visual impairments. Natural language processing dates back to the beginnings of machine learning in the 1950s, when Alan Turing wrote a seminal paper called "Computing machinery and intelligence" [38]. In this paper, Turing proposes an imitation game where an interrogator needs to figure out the gender of a man and a woman invisible to him, by asking them questions. If one of the invisible persons is replaced by a computer and the interrogator cannot determine anymore if he is talking to the human or the computer, the computer can generate human cognitive capabilities. This is now called the Turing test.

There is a lot of interest in natural language processing in research and practice since it simplifies human-machine interaction, however there are still limits to what natural language processing can do today. Machines are capable of understanding spoken commands or determine the subject of a text, but, to humans, a sentence or text has meaning, to a machine it is an array of characters. A machine does not understand the context what makes word sense disambiguation a tricky task. The word "meeting" can be a noun, as in "we are in a meeting", or a verb, as in "we are meeting an my office".

10.3.1 Text mining

Text mining or text analysis, aims to extract information from texts automatically. The applications of text mining are innumerable. Text mining is used for text categorization, text clustering, document summarization or opinion mining, but many other use cases exist. We have already covered spam filtering, which is an application of text classification. Emails are classified into legitimate and into spam mails.

Sentiment analysis

Sentiment analysis, also called opinion mining, is the field of study that analyzes peoples opinions, sentiments, evaluations, appraisals, attitudes, and emotions towards entities such as products, services, organizations, individuals, issues, events, topics, and their attributes [24]. Sentiment analysis is extensively used by companies to determine the popularity of products or services in order to improve them. It is also used in politics to make predictions on the outcome of elections or to determine the popularity of TV programs or series. With the advent of Social Media, such as Twitter or Facebook, it has become very easy for anybody to post an opinion on a movie or on a new product on the internet. Analyzing Social Media has, thus, become mainstream in many areas. Social media mining is used for personalized ads or for targeted marketing campaigns. It has also been used to determine the credit worthiness of a person. If an applicant for a load or mortgage has wealthy Facebook friends, he is likely to be wealthy himself and considered credit worthy.

Sentiment analysis uses natural language processing methods to determine the opinions or sentiments in a text. We have already seen the bag-of-words approach. Sentiment words, such as "great", "excellent" and "poor" are counted. If a text contains many positive sentiment words or more positive than negative sentiment words, it is considered a positive review. However, the bag-of-words approach considers only individual words and cannot handle sentiment polarity shifters. A sentiment polarity shifter changes the meaning of an expression to the opposite. For instance "the new Tesla is great" is a positive statement, "the new Tesla is not great" is negative. Here the word "not" acts as a polarity shifter. We have already seen the bag-of-words approach in section 4.1.2, used, for instance, in naïve Bayes spam filtering. We have also seen a more sophisticated approach that uses n-grams to capture the relationship between words. A tokenizer is used to split a text into sequences of words. If we have sequences of two words, they are called bigrams, and we analyze how often a word X is followed by word Y. we can determine the relationship between them. One approach is now to count bigrams and do frequency analysis on the bigrams. If we have many bigrams such as "not great" or "very bad" the text is considered a negative review.

Neural networks and, in particular, deep learners have been used for sentiment analysis. One big advantage of deep learners is their capability to automatically extract features as we have seen in Chapter 8. When using a shallow learner, such as the multilayer perceptron, we have to extract the features manually. A feature might be a part-of-speech tag on a word. In part-of-speech tagging, we assign a word in a text to a particular part of speech based on its relationship with adjacent and related words in a text. A word is identified as noun, verb, adjective, etc., for word-category disambiguation. For instance, the word "fly" can mean the insect, when it is a noun, or "to fly", when it is a verb. After tokenization and part-of-speech tagging, the word stem, if found, and the words are grouped according to their word stems, a process called lemmatization. The words "had", "has" and "having" are grouped since they have the same word stem, "have". Lemmas extend the feature set by adding word groups based on stems. Using deep learners, these steps, such as lemmatization and part-of-speech tagging, are done automatically, which is a big advantage over shallow approaches. However, deep learners require a large amount of training data that is not always available.

10.4 Other applications

Machine learning has many more applications that cannot all be covered in one chapter or book, but some notable area of machine learning utilization is the military. Drone surveillance produces a tsunami of data that needs to be analyzed quickly before, for instance, a potential enemy has disappeared. The Department of Defence has specifically established an Algorithmic Warfare Cross-Functional Team (AWCFT) in order to integrate artificial intelligence and machine learning more effectively across operations so as to maintain advantages over increasingly capable adversaries and competitors [42]. Using AI, Project Maven will initially provide vision algorithms for object detection, classification, and alerts, according to the memo establishing the team. Google has supported the Pentagon with a special API to its Tensor-Flow machine learning framework for analyzing images from drones. Google is providing AI to the Pentagon Project Maven, which aims to speed up analysis of drone footage by automatically classifying images of objects and people in the MQ-9 Reaper drone. Its goal is to make the drone attack military targets automatically. Ultimately, the goal is to develop autonomous systems, such as drones and other airborne tactical areal systems, that can attack independently.

Machine learning can be used to analyze and respond to cyber attacks as described before. Machine learning has been used in cyber security for detecting anomalies in network traffic and malware. However, machine learning can also be used to create malware. Developing malware, Trojans, ransomeware or malicious scripts, is largely a manual process for a cyber criminal. Machine learning can automate this process or speed it up considerably. Algorithms based on generative adversarial network (GAN) have already been proposed [12]. Machine learning can generate malware which is not detectable by traditional intrusion detection systems. Malware authors usually have no access to the detailed structures and parameters of the machine learning models used by malware detection systems, therefore, they can only perform black-box attacks [12]. Using machine learning, they can modify code on the fly and adapt to detection systems and, thus, form polymorphic malware. They can also poison the machine learning engine of the detection system, like viruses already do for AntiVirus software. More sophisticated attacks use botnets and swarm intelligence. A swarm is a decentralized, self-organizing system. Bots that are interconnected through the Internet form a botnet. The community of computers in a botnet transmit spam or viruses across the web without their owners permission [43]. Self-learning swarmbots form a hivenet. Hivenets share information to create a customized attack. They can, for instance, attack any access point in a company network. Bots can run on any device connected to the internet, smart phones, baby phones, TV boxes or soundbars.

The Internet of Things (IoT) is an evolution of the Internet where everyday objects such as refrigerators, sensors or actuators are connected to the Internet and can communicate with other devices. Internet of Things is the connectivity of the Internet into the physical world. Technological advancements lead to smart objects being capable of identifying, locating, sensing and connecting and, thus, leading to new forms of communication between people and things and things themselves [6]. Internet of Things produces a lot of data and is only useful when the data can be analyzed effectively. Internet of Things has given machine learning a whole new application field. The data produced by Internet of Things enabled devices can be of any form; numeric data, streaming data, text data or binary data. The data is transmitted to the cloud, where it is being analyzed and visualized for end users to be assessed. Cloud computing has become a de facto platform to enable content delivery to consumers [9]. Internet of Things can connect all equipment in a whole production plant to monitor its health state and determine if a device or part needs maintenance or replacement. Often, it is not feasible to transmit all data into the cloud due to limited bandwidth or storage in the cloud. The premise of seamless integration of all Internet of Things enabling technologies, along with advancements in IPv6 and semantic services, positioned the industry in a premature state [29]. Major advancements in wireless technology, Radio-frequency identification (RFID) and improvements in battery technology enabled Internet of Things, but effectively analyzing the data produced is still challenging. Often, the data is pre-processed and analyzed on systems close to the physical devices and only a consolidated data set is sent to the cloud. This is called edge computing. Data analysis software is deployed on the edge computer. Edge computers are machine learning enabled and usually support deep learning too. The edge computer monitors the connected devices for potential problems using predictive and preventive maintenance as described in section 10.1.2. If it determines that an equipment needs maintenance, it sends an alert to the cloud where a service technician can be assigned to correct the problem. Often, several edge computers are deployed, each monitoring and analyzing different equipment. The data mining system in the cloud analyses the consolidated information transmitted from the different edge computers. This is called distributed analytics. Distributed analytics spreads the workload aver multiple nodes, the edge computers. This is called fog computing or fog networking, where a network of connected edge computers carry a substantial amount of computing and storage. This results in faster insights into the health state of the connected devices. Ideally, the edge computers can communicate with each other and do collaborative machine learning, see section 11.1. However, current Internet of Things solutions are not as advanced yet.

Machine learning enabled Internet of Things solutions are not limited to predictive and preventive maintenance. They can be used to optimize the production in a factory, save resources such as energy or raw materials and reduce waste.

Chapter 11

Future Development

CONTENTS

There is no agreed upon definition of artificial intelligence in literature. But generally speaking, artificial intelligence describes tasks that are normally done by humans on a computer. However, the way machines undertake human tasks is fundamentally different from how humans do them, humans are still superior in many of them. Whether it is a child learning to distinguish cats from dogs or old people learning the rules of a card game, humans have an unparalleled capability for learning. A child sees a cat once and then knows what a cat is, independent of its size or fur color. A machine learning scheme needs to be trained on thousands of images of cats in different positions and at different ages to be able to recognize a cat with a different fur pattern. Humans are excellent at classifying things we see into categories. This is essential for our survival. Primordial human beings had to be able to quickly distinguish foe from prey when hunting. Similarly, we need to be able to distinguish potentially dangerous traffic situations from normal ones. Since we constantly encounter new traffic situations, we have to be able to quickly assess if they are potentially dangerous or not. We see many vehicles on the street that we ignore, but suddenly all our attention is focused on the one that makes a strange maneuver. It is largely unknown today how humans subconsciously decide what is relevant. This is one of the big challenges autonomously driving cars have to solve.

Over the past few decades, a lot of progress has been made in unraveling how human learning works at the molecular level. In recent years, neuroscience has gained a relatively deep understanding of how memories are formed. Learning and memory formation are linked to the strengthening and weakening of connections among brain cells. Neurons that frequently interact form a bond that let them transmit information more easily and accurately between eachother. In stark contrast, is our limited understanding of how these same memories are maintained [27].

How we recall episodes from the past and replay them like a movie in our head is largely unknown. The ability to encode and retrieve our daily personal experiences, called episodic memory, is supported by the circuitry of the medial temporal lobe (MTL), including the hippocampus, which interacts extensively with a number of specific distributed cortical and subcortical structures [5]. There are three areas in the human brain involved in explicit memory storage: The hippocampus, the neo-cortex and the amygdala. How the memories are stored is not well understood today. Some studies have shown that learning modifies the DNA [27]. Epigenetic mechanisms probably have an important role in synaptic plasticity and memory formation [20]. The DNA is the most stable memory location in the human body and would, thus, be a logical place to retain information, but this is pure speculation. Humans also forget things. The capability to forget is to protect us from an information overflow. The memory is divided into three stores, the sensory memory, that stores information from our senses, the short-term memory (STM) and the long-term memory (LTM). The sensory memory is extremely short, it retains information only for a few seconds. The STM, as the name suggests, stores information for a short period of time, usually between 20 and 30 seconds. If the information is rehearsed, it is retained longer and might become a LTM. How the brain filters the information that it retains and how it decides what information it forgets is not well understood. We might remember an argument from last year, but only vaguely remember a conversation we had a couple of weeks ago. Here, again, we do not know what the brain considers relevant to retain and what it forgets. A heated argument is probably more important to store since it might have greater consequences than a conversation about the weather, but how this filter works, we do not know.

These unknowns are a big challenge in artificial intelligence and in machine learning. How can we imitate something on the computer when we do not know how it works in real life? Even if we are still in the dark about many aspects of how learning and memorizing works in humans, when trying to imitate it, we learn what problems nature had to solve to give us the capabilities we have. Learning by imitating a biological mechanism helps us better understand the problem at hand, even if the solution we find to solve it is very different from the solution in nature. To advance artificial intelligence and machine learning we need both, a better understanding of the biological mechanisms and improved implementations that try to imitate them. Machine learning is an interdisciplinary field of study and it is expected that we will see more researchers cooperating with each other to bring together the knowledge of different areas. Linguists will collaborate with data scientists to improve the text mining capabilities, neurobiologists will collaborate with computer scientists and roboticists to advance the skills of robots and autonomous systems. A lot of effort has already been undertaken in the industry and in academia since improved systems help us to make production sites or traffic more efficient and reduce energy consumption and emissions. Machine learning can potentially reduce health services and elderly care costs, improve safety measures oin the roads and in the work space, to name a few application areas. We will also likely see new application fields that we might not have thought of yet.

11.1   Research directions

Currently, a machine learning algorithm is trained for one thing, for instance, to translate text from one language into another or to steer a robot or an autonomous car. This is in contrast to human intelligence.

Humans can read and write, make calculations, play football, sing and play the piano. A lot of research is performed to develop algorithms that are capable of learning different tasks. This research area is called artificial general intelligence (AGI), contrary to artificial narrow intelligence (ANI). AGI has the aim to perform any intellectual task a human can do. Instead of developing one single system that can learn anything, we will probably see several systems working together for collaborative learning. There are already systems available where several learners can be deployed, for instance, one that is trained to translate English into Russian and another one trained to translate English to German. The advantage of these machine learning engines is that they can be modularly built and enhanced, for instance, with new languages or with new capabilities, such as text to speech functionality. Modular systems are easier to maintain and test.

The Internet of Things (IoT) has given machine learning a whole new application field. IoT creates a tsunami of heterogeneous data in different formats at high velocity. IoT systems are also dynamic since new sensors can be added or removed, whole buildings can be interconnected for smart building control and whole production sites are wired. A machine learning system needs, thus, to be enhanced as well in order to analyze the data of a newly connected robot or security system. Instead of retraining a machine learning scheme with data from newly connected systems, the machine learning engine can be enhanced with a new module, but a newly connected production site might already have a data mining system in place. The data mining system has to be integrated in an existing solution that might use a different technology. Machine learning engines will need integration capabilities for collaborative learning using multiple technologies. A better solution might be a system for continuous learning. Instead of training, testing and deploying a machine learning engine, it can continuously learn from new data without retraining and redeployment. Such a system needs to automatically perform the pre-processing, feature extraction, learning and evaluating tasks as soon as new types of data arrive. Such a system is called an automated machine learning system. An automated machine learning system performs the end-to-end process of data pre-processing untill analysis, automatically, without human intervention. Fully automated machine learning systems do not exist yet but automated machine learning is an active area of research for obvious reasons. They are cheaper if they perform laborious tasks themselves and they are faster. Unsupervised learning will probably be the method of choice for automated machine learning since no labeled data is required. Also, in nature, most learning is unsupervised. Most of the learning in the future will probably be continuous, unsupervised learning.

The recent progress we have seen in machine learning is owed, among other, to advances in hardware developments. In particular, the development of GPUs has provided the computing power to execute machine learning algorithms efficiently. Today, chips are developed specifically for machine learning and artificial intelligence. The next evolution in hardware development will be most likely be quantum computers. Quantum computing considerably speeds up computing operations and will increase the velocity of training a learning algorithm and decrease the execution time of machine learning algorithms by magnitudes.

The black box approach of trained machine learning systems has caused unease in society and scenarios such as a robot who suddenly attacks humans or an autonomously driving car causing an accident without an understanding of who is liable are circulating in the press. Many of those fears are currently unfounded. For one, there are currently no such advanced systems available, and secondly, giving total control to a machine is not desirable. No company wants to leave its security in the hands of a machine. No physician will let a machine diagnose his patients. Machine learning will support many specialists, for instance, in detecting cyber security breaches or anomalies in medical images, but an Army general deploying uncontrollable robots in warfare is an unlikely scenario. Even if machine learning engines are making decisions by themselves, the ultimate choice of what should be done will still remain in human hands as long as machines cannot clearly justify and explain their actions.

Machine learning schemes are capable of determining the topic of a text or an anomaly in sensor data, but they do not understand the context of the text or why a certain sensor value is abnormal. More research is required for a machine learning engine to be capable of understanding context and meaning. When a learner has determined that a book is about the second world war, it still does not understand its context in human history. If an anomaly has been detected by a machine learning scheme, it does not understand the consequence of the anomaly. The artificial intelligence system should be able to explain its conclusions, ultimately, it should be able to reason. However, at the current rate of research and development, there is still a long way to go until we will see such advanced systems.

References

[1] Abbott, D. 2014. Applied Predictive Analytics: Principles and Techniques for the Professional Data Analyst. John Wiley & Sons.

[2] Buczak, A.L. and Guven, E. 2016. A survey of data mining and machine learning methods for cyber security intrusion detection. IEEE Communications Surveys & Tutorials, vol. 18, no. 2, pp. 1153–76.

[3] Bzdok, D., Krzywinski, M. and Altman, N. 2018. Machine learning: Supervised methods. Nature Methods, vol. 15, p. 5.

[4] Chapelle, O., Schoelkopf, B. and Zien, A. 2006. Semi-Supervised Learning, The MIT Press, Cambridge, Massachusetts.

[5] Dickerson, B.C. and Eichenbaum, H. 2010. The episodic memory system: Neurocircuitry and disorders. Neuropsychopharmacology, vol. 35, no. 1, pp. 86–104, "http://www.ncbi.nlm.nih.gov/pmc/articles/PMC2882963/".

[6] Dohr, A., Modre-Opsrian, R., Drobics, M., Hayn, D. and Schreier, G. 2010. The internet of things for ambient assisted living. Information Technology: New Generations (ITNG), 2010 Seventh International Conference on, pp. 804–9.

[7] Glorot, X., Bordes, A. and Bengio, Y. 2011. Deep sparse rectifier neural networks. In Proceedings of the Fourteenth International Conference on Artificial Intelligence and Statistics, G. Geoffrey et al. (eds.). PMLR, Proceedings of Machine Learning Research, pp. 315–23.

[8] Goodfellow, I., Bengio, Y. and Courville, A. 2016. Deep Learning, MIT press.

[9] Guelzim, T. and Obaidat, M.S. 2016. Chapter 12—Cloud computing systems for smart cities and homes. In Smart Cities and Homes, Morgan Kaufmann, Boston, pp. 241–60.

[10] Han, J., Pei, J. and Kamber, M. 2011. Data Mining: Concepts and Techniques, Elsevier.

[11] Hastie, T., Tibshirani, R. and Friedman, J. 2009. The Elements of Statistical Learning: Data Mining, Inference, and Prediction, Second Edition, Springer Series in Statistics.

[12] Hu, W. and Tan, Y. 2017. Generating Adversarial Malware Examples for Black-Box Attacks Based on GAN, arXiv:1702.05983.

[13] Hopfield, J. 1982. Neural Networks and Physical Systems with Emergent Collective Computational Abilities, vol. 79.

[14] James, G., Witten, D., Hastie, T. and Tibshirani, R. 2013. An Introduction to Statistical Learning, Springer, New York, NY.

[15] Koza, J.R., Bennett, F.H., Andre, D. and Keane, M.A. 1996. Automated design of both the topology and sizing of analog electrical circuits using genetic programming. In J.S. Gero and F. Sudweeks (eds.). Artificial Intelligence in Design 96, Springer Netherlands, Dordrecht, pp. 151–70.

[16] Krizhevsky, A., Ilya, S. and Hinton, G.E. 2012. ImageNet Classification with Deep Convolutional Neural Networks, pp. 1097–105.

[17] Larose, D.T. and Larose, C.D. 2015. Data Mining and Predictive Analytics, Wiley Publishing.

[18] LeCun, Y., Bengio, Y. and Hinton, G. 2015. Deep learning. Nature, vol. 521, no. 7553, pp. 436–44.

[19] Lecun, Y., Bottou, L., Bengio, Y. and Haffner, P. 1998. Gradient-based learning applied to document recognition. Proceedings of the IEEE, vol. 86, no. 11, pp. 2278–324.

[20] Levenson, J.M. and Sweatt, J.D. 2005. Epigenetic mechanisms in memory formation. Nature Reviews Neuroscience, vol. 6, no. 2, p. 108.

[21] Lillicrap, T.P. and Santoro, A. 2019. Backpropagation through time and the brain. Current Opinion in Neurobiology, vol. 55, pp. 82–9.

[22] Lipton, Z., Berkowitz, J. and Elkan, C. 2015. A Critical Review of Recurrent Neural Networks for Sequence Learning, 2015, arXiv:1506.00019.

[23] Liu, B. 2011. Web Data Mining: Exploring Hyperlinks, Contents, and Usage Data, 2 edn, Springer, Heidelberg.

[24] Liu, B. 2012. Sentiment Analysis and Opinion Mining, Morgan & Claypool.

[25] Marsland, S. 2014. Machine Learning: An Algorithmic Perspective, Second Edition, Chapman & Hall/CRC.

[26] McCulloch, W.S. and Pitts, W. 1943. A logical calculus of ideas immanent in nervous activity. Bulletin of Mathematical Biophysics, no. 5, pp. 115-133.

[27] Miller, C.A., Gavin, C.F., White, J.A., Parrish, R.R., Honasoge, A., Yancey, C.R., Rivera, I.M., Rubio, M.D., Rumbaugh, G. and Sweatt, J.D. 2010. Cortical DNA methylation maintains remote memory. Nature Neuroscience, vol. 13, no. 6, pp. 664–6, "http://www.ncbi.nlm.nih.gov/pmc/articles/PMC3043549/".

[28] Ormond, J. 2018. Fathers of the Deep Learning Revolution Receive ACM A.M. Turing Award. ACM A. M. Turing Award.

[29] Oteafy, S.M.A. and Hassanein, H.S. 2012. Resource re-use in wireless sensor networks: Realizing a synergetic internet of things. Journal of Communications, vol. 7, no. 7.

[30] Peng, F., Schuurmans, D. and Wang, S. 2004. Augmenting naive bayes classifiers with statistical language models. Information Retrieval, vol. 7, no. 3, pp. 317-4.

[31] Poole, D.L. and Mackworth, A.K. 2017. Artificial Intelligence: Foundations of Computational Agents, 2 edn, Cambridge University Press.

[32] Purves, D., Augustine, G.J., Fitzpatrick, D. et al. 2001. Generation of Neurons in the Adult Brain. Neuroscience 2nd edition, Sinauer Associates, 2001.

[33] Rav, D., Wong, C., Deligianni, F., Berthelot, M., Andreu-Perez, J., Lo, B. and Yang, G.Z. 2017. Deep learning for health informatics. IEEE Journal of Biomedical and Health Informatics, vol. 21, no. 1, pp. 4–21.

[34] Rosenblatt, F. 1958. The perceptron: A probabilistic model for information storage and organization in the brain. Psychological Review, no. 65, pp. 386-408.

[35] Samuel, A.L. 1959. Some studies in machine learning using the game of checkers. IBM J. Res. Dev., vol. 3, no. 3, pp. 210–29.

[36] Siegel, E. 2013. Predictive Analytics: The Power to Predict Who Will Click, Buy, Lie, or Die, Wiley Publishing.

[37] Tan, P.N., Steinbach, M., Karpatne, A. and Kumar, V. 2013. Introduction to Data Mining, Pearson.

[38] Turing, A.M. 1950. Computing machinery and intelligence. Mind, vol. LIX, no. 236, pp. 433–60.

[39] Whang, S.E., McAuley, J. and Garcia-Molina, H. 2002. Compare Me Maybe: Crowd Entity Resolution Interfaces, Stanford University.

[40] Witten, I.H., Frank, E., Hall, M.A. and Pal, C.J. 2016. Data Mining, Fourth Edition: Practical Machine Learning Tools and Techniques, Morgan Kaufmann Publishers Inc.

[41] Wlodarczak, P., Soar, J. and Ally, M. 2015. Multimedia data mining using deep learning. Digital Information Processing and Communications (ICDIPC), 2015 Fifth International Conference on, IEEE Xplore, Sierre, pp. 190–6.

[42] Work, B. 2017. Establishment of an Algorithmic Warfare Cross-Functional Team (Project Maven), Department of Defense.

[43] Zafarani, R., Abbasi, M.A. and Liu, H. 2014. Social Media Mining, Cambridge University Press, Cambridge.

Index